7 REASONS TO BELIEVE

COMPELLING EVIDENCE FOR THE EXISTENCE OF GOD

KEVIN SIMINGTON

SMART FAITH PRESS

7 REASONS TO BELIEVE
Compelling Evidence for the Existence of God

© Copyright 2020 Kevin Simington

All rights reserved. No part of this publication may be reproduced, stored in a retrieval system or transmitted in any form by any means electronic, mechanical, photocopying, recording or otherwise, without the prior written permission of the author.

Unless otherwise specified, Scripture quotations are from the New International Version Bible, copyright © 1973. 1978, 1984, 2011, Zondervan, Grand Rapids, Michigan, USA.

EDITORS: Tony Baker, Sandra Simington

❦ Created with Vellum

CONTENTS

Preface   v

1. The Nature And Limitations Of Evidence   1
2. Removing The Elephant   13
3. Evidence 1: The Origin Of The Universe   23
4. Evidence 2: The Origin Of Life   41
5. Evidence 3: Intelligent Design   63
6. Evidence 4: The Existence Of Objective Moral Values   95
7. Evidence 5: God's Intervention In Human History   111
8. Evidence 6: The Resurrection Of Jesus Christ   131
9. Evidence 7: Personal Experience   147
10. Pascal's Wager   165

Other Titles By Kevin Simington   181
Finding God When He Seems To Be Hiding   183
Making Sense of the Bible   185
No More Monkey Business: 187
Rethinking The Gospel   189
The Little Book of Church Leadership   191
Welcome To The Universe   193
Someone Else's Life   195
The Starpath Series   197
Connect With Kevin Simington   199
About the Author   201
Notes   203

PREFACE

This is a book I have wanted to write for a long time. It examines the seven most persuasive pieces of evidence for the existence of God. While it is primarily aimed at people who are not yet Christians, it will also help those who already believe, to strengthen their own faith and equip them to more confidently share that faith with others.

The evidence presented in this book is drawn from the areas of molecular biology, genetics, cosmology, philosophy, history and personal experience. The evidence that is science-based is necessarily technical (although it will be explained for the average reader). There would be no point trying to summarise or over-simplify the technical evidence, because to do so would detract from its compelling nature and open the door for intellectual sceptics to label the arguments as shallow and naïve.

Portions of the evidence that I present in this book have also appeared in slightly different forms in some of my other books. For instance, elements of the scientific arguments have appeared in *No More Monkey Business: Evolution in Crisis*. Similarly, some of the points I make in the chapter on historical evidence are echoes of points made in my first book, *Finding God When He Seems to Be Hiding*.

But most of the book is brand new material, and even those arguments that have appeared in other books have been significantly updated.

In many ways, this book is the sequel to *Finding God When He Seems to Be Hiding*. That was my first book and it dealt with answering the questions and objections **against** Christianity that are commonly posed by sceptics. It was dealing with the **negative** roadblocks that stop many people even considering Christianity. This book deals with the **positive** evidence *for* Christianity. The two books are perfect companions and together form a powerful defence of the Christian faith. In a sense, this book picks up where *Finding God* left off. It effectively says, *"Now that you've had your objections answered and you are ready to consider Christianity with an open mind, here is the compelling evidence for the existence of God and the truth of the Christian message."*

It is my hope and prayer that *7 Reasons to Believe* will be a great blessing to you and to all who seek the truth.

1
THE NATURE AND LIMITATIONS OF EVIDENCE

I'd like to prove to you that God exists. But I can't.

No one can.

Don't get me wrong, there is plenty of evidence that points us to the existence of God: evidence that, in my opinion, is undeniable and utterly convincing. But absolute PROOF is simply not possible. That's because of the nature of proof itself. Let me explain.

Philosophically speaking, it's impossible to unequivocally prove anything. In fact, I can't prove that I exist. I can't even prove that the physical universe exists; that it is not some giant simulation into which I am plugged. For the most part, we must satisfy ourselves not with indisputable proof, but with the weight of reasonable evidence which leads us to conclude that something is *probably* true. Thus, the weight of reasonable evidence leads me to conclude that I probably exist and that the physical universe is probably real. The same is true in regard to God. While I cannot indisputably prove his existence, the weight of observable evidence leads me to conclude that the most likely explanation of that evidence is the existence of an all-powerful Creator God.

In this book, I will present the seven most convincing areas of evidence that indicate the existence of a Creator God. As I do so, I do not presume to know your current view on the matter. Indeed, it is likely that a wide range of viewpoints will be represented among those who read this book:

- **Atheists:** those who have already formed a strong opinion that there is no God.
- **Agnostics:** those who are unsure or who believe it is impossible to be certain about God's existence.
- **Disinterested:** those who are largely indifferent to the question of God's existence or who may regard the issue as irrelevant to life.
- **Sympathisers:** those who have a sneaking suspicion that God exists, but who have not investigated the matter at any significant depth.
- **Seekers:** those who have a strong desire to settle the question of God's existence and may have unresolved questions or doubts.
- **Supporters:** those who already believe in God and are wanting to investigate the evidence further in order to establish a firmer base for that belief.

In presenting the evidence for God's existence I will not be assuming that every reader is in the same 'category'. If I am to be effective, however, I will have to regularly address the most extreme of the negative positions: atheism. In doing so, I am not inferring that everyone reading this book is an atheist, but if I don't address the questions and objections commonly raised by atheists, the arguments presented in this book will fall short of being comprehensive and 'storm proof'.

Consequently, as each major area of evidence is outlined and investigated throughout this book, the corresponding counter-arguments and objections commonly raised by atheists must also be discussed. The possible existence of God continues to be a hotly debated topic in philosophical circles, and if this book is to be worthwhile it must accurately reflect that debate, including its intensity. In doing so, this will sometimes necessitate forcefully worded arguments. If, at certain points throughout the book, you think that I am addressing objections that you are not raising, or dismantling arguments you are not making, please don't feel inappropriately besieged or embattled. I am simply trying to address all possible objections from a wide variety of readers.

Intelligent atheists are, of course, familiar with many of the arguments discussed in this book and, in some instances, have fought for decades to either dismiss or devalue them. As an interested observer in this ongoing debate and an active participant myself, I hope to bring an informed, up-to-date evaluation of current arguments.

A fascinating trend that I have witnessed in recent years has been the erosion of the evidence-base for atheism. As scientific knowledge in various fields has continued to develop, the evidence for the existence of a supernatural Creator has dramatically increased rather than diminished. Science, once the hoped-for validator of atheism, has turned out to be a major thorn in its side. As you will discover in the course of this book, remarkable developments in various scientific fields have increasingly revealed the impossibility of this universe having created itself. Modern science has become the friend of the theist (the believer in God) and has left atheists scrambling with increasingly imaginative and unlikely theories to explain the formation of the universe and the development of complex, intelligent life.

Reflecting on these imaginative theories, Dr Christopher J. Isham, Britain's leading quantum cosmologist, and an astrophysicist at Imperial College of London, recently wrote:

> "The idea that the Big Bang supports theism [belief in God] is greeted with obvious unease by atheist physicists. At times this has led to <u>wild scientific theories</u> being advanced with a tenacity which so exceeds their intrinsic worth that one can only suspect the operation of psychological forces lying very much deeper than the usual academic desire of a theorist to support his or her theory."[1] (Emphasis mine).

Science is not the atheist's only problem. In the course of this book, you will also see how philosophy, history and personal experience add further powerful evidence for the existence of a supernatural Creator.

You may ask, 'If this evidence is so convincing, why aren't more people believers?'

Great question! I'm glad you asked it!

The problem lies with our human predisposition to believe what we **want** to believe and to view any new or contradictory evidence through the lens of what we **already** believe. If we hold strongly to a particular viewpoint and someone presents evidence that conflicts with it, our automatic response is often to attempt to refute that evidence or reinterpret it in a way that allows us to maintain our current belief. In other words, our pre-existing bias makes it extremely difficult for us to be open-minded and evaluate new evidence dispassionately. The more passionately we hold to a certain viewpoint, the stronger will be our interpretive bias and the less open we tend to be in considering new evidence at face value.

This is true for all of us. It is true for me as a Christian. When I come across a piece of literature that claims to disprove some of my beliefs, my automatic reaction is to read it with a combative mindset. It takes a strong act of my will to lay aside my critical spirit and evaluate the evidence and the arguments with an open mind.

Atheists and sceptics often have the same problem. Those who hold an atheistic worldview often tend to be very passionate about it and are extremely reluctant to consider the alternative. They *want* to believe that there is no God, because the alternative – the existence of an all-powerful Creator God to whom they might be ultimately accountable – would be extremely inconvenient. It would ruin their whole self-directed philosophy and require a monumental change of life. Thus, atheists have a strong vested interest in maintaining their disbelief in God, and this is why they often find it very difficult to consider contrary evidence with dispassionate evaluation. In some cases, the most outspoken atheists are already utterly convinced of their worldview, and no amount of evidence to the contrary will change that.

Have I just described you? Have you already made up your mind? Maybe you are reading this book reluctantly, already summoning your counter-arguments and readying yourself to rip these arguments to pieces. (I reiterate, I don't assume that every reader is in this position, but I must address this most extreme position).

Can I ask you to do something? Can I ask you to attempt to lay aside your preconceptions and presuppositions and consider the evidence in this book with an open mind? I am not asking you to read this book unthinkingly. By all means, carefully analyse and evaluate everything. Hold the evidence up to careful, rigorous scrutiny. But try to do so as an honest seeker of the truth, rather than a crusading destroyer of alternate views. Try to approach these arguments and pieces of evidence with a magnifying glass rather than a flame thrower.

This, of course, is a difficult thing to do - to suspend your personal biases and presuppositions and open-mindedly follow the evidence wherever it leads you.

I recently read the now-famous book that sent shock-waves around the world when it was published in 2007; *There is a God: How the World's Most Notorious Atheist Changed His Mind*. The book is the self-reflections of Antony Flew, arguably the most influential atheist of the 20th century, who eventually came to believe in God after a lifetime studying science and philosophy.

The claim of the book's subtitle, *"The World's Most Notorious Atheist"*, is not an exaggeration. Antony Flew was, for half a century, an intellectual giant among the world's atheists and their most influential voice. His 1950 publication, *Theology and Falsification*, became the most widely reprinted philosophical publication of the last 70 years and established a systematic philosophical foundation for modern atheism. Over the following 50 years, he wrote about 30 books and papers, including *God and Philosophy, The Presumption of Atheism* and *How to Think Straight*, all of which became the mainstays of the worldwide atheist movement. You may not have heard of Antony Flew, but he is without doubt the most influential atheist of at least the last century. The more recently well-known Richard Dawkins and Christopher Hitchens became household names due to the rise of the Internet, but they and others like them are not even on the same page as Antony Flew, whose towering intellect forged the modern atheist manifesto upon which all others have subsequently built.

Flew's public rejection of atheism and his conversion to theism (belief in a Creator God) in 2007 was, therefore, a cataclysmic event within the atheist community. The atheist world was in uproar and scrambled to make sense of his capitulation, even resorting to inferences that, perhaps, he was suffering from dementia. His book, however, leaves us in no doubt as to the continued sharpness and perspicacity of his mind.

So, what changed his mind – a mind that for so long had seemed resolutely opposed to any idea of the supernatural? Did he have some kind of religious experience? A divinely inspired epiphany? Did God appear to him in a 'burning bush' type encounter?

No. Flew is very clear about this. He says that it was not a religious experience of any kind that changed his mind but, rather, a methodical, dispassionate, incremental analysis of the scientific evidence over many years that inexorably led him to the conclusion that there must be a God. He says that he decided to lay aside his preconceptions and personal bias and doggedly follow the evidence wherever it led him, no matter how uncomfortable that destination was to his beliefs.

And what was the evidence that he followed? What did Antony Flew find so convincing? He specifies three areas of evidence that he found utterly compelling. I will let him describe them to you in his own words:

> "I now believe that the universe was brought into existence by an infinite Intelligence ... Why do I believe this, given that I expounded and defended atheism for more than half a century? ... Science spotlights three dimensions of nature that point to God. The first is the fact that nature obeys laws. The second is the dimension of life, intelligently organized and purpose-driven beings, which arose from matter. The third is the very existence of nature itself."[2]

Flew explains what he means regarding these three areas by posing three questions:

> "How did the laws of nature come to be? ... How did life originate from non-life? And the third is the [biggest] problem [facing] cosmologists: How did the universe, by which we mean all that is physical, come into existence?"[3]

Flew contends that atheism and evolutionary theory have no plausible answers to these big questions. In fact, he argues that the more evidence that science uncovers regarding the extraordinary complexity of the universe, the more it points to a supernatural creative origin. This evidence includes the irreducible complexity of the smallest biological elements, the obvious signs of intelligent design throughout nature, the incredible and statistically impossible fine tuning of the laws of nature, the impossibility of non-living matter giving rise to life, and the ultimate question of the origin of the universe itself.

It doesn't matter if you don't yet have a comprehensive understanding of these concepts; they and other compelling evidence will be explained thoroughly in the course of this book. For the moment, it is sufficient for you to simply understand the importance of open-minded enquiry. Dr Flew's conversion to belief in God demonstrates the fact that there is very convincing evidence for God's existence for those who are willing to consider it. Flew insists that *"we must follow the argument wherever it leads"* and states that, in his case, it led him to finally concede the existence of a supernatural Creator.

It is no small thing for so prominent an atheist to publicly repudiate his life's work and admit that he was wrong. In doing so, he states that the process of his conversion was not easy because he, like most atheists, was *"dogmatic"* in his assertion that there is no God.[4] He, therefore, sympathises with atheists who have closed their minds to any possibility that they may be wrong.

In his chapter entitled *"A Pilgrimage of Reason"*, Flew poses a challenge to atheists whose minds are already made up:

> "I therefore put to my former fellow-atheists the simple central question: 'What would have to occur, or to have occurred, to constitute for you a reason to at least consider the existence of a superior Mind?'"[5]

It is a good question. It is a question that everyone should consider. How strong would the evidence need to be for you to consider that there may be a God?

Please allow me to make a personal appeal at this point. Before you leave this world behind and venture through the final curtain of death, wouldn't it be wise to fact-check your belief (or unbelief) against the hard evidence? As Flew says, *"we must follow the argument wherever it leads."* Surely, you have nothing to lose in doing so. If the evidence confirms your unbelief, then you have lost nothing. But if the evidence indicates the existence of a Supreme Being in whose presence you may one day stand, surely it is prudent to find that out NOW so that you can do something about it. It seems to me that to go through life determined to disbelieve in God and refusing to even consider the evidence is a very risky philosophy. You are risking everything on the bet that there is no God.

THE MOST IMPORTANT QUESTION IN THE WORLD

This brings me to a final preliminary issue that needs to be addressed. There are many people who regard the question of God's existence as unimportant and irrelevant to their life.

They may either have a vague belief in God's existence or perhaps have reached a superficial conclusion that he probably doesn't exist. Either way, they are not interested in pursuing the question any further because they see it as a nebulous or unanswerable philosophical question that has little relevance to their everyday life.

Perhaps that is your attitude.

If that is so, let me point out that the question of God's existence is the most important question in all of life. It is more important than the questions of who you will marry, where you will live, what career you will choose and all the other 'big' decisions of life. The 'God question' is the biggest of them all. In fact, it's bigger than all of the other questions *combined*, because the answer has the potential to change EVERYTHING. The answer to the 'God-question' potentially has eternal consequences.

If God **does not** exist, then we are just the products of biological chance. It would mean that this life is all there is: there is no life after death, no judgment, no final accountability to an ultimate law-giver and moral ruler. It would mean that we are the sovereign rulers of our own lives, masters of our own destiny, free agents, able to choose our own moral standards and map our own course without reference to a higher power. If this is the case, then the old adage, *'eat, drink and be merry, for tomorrow we die'* makes perfect sense. If there is no God, then we may as well suck the marrow out of life, make pleasure our ultimate goal and pursue it relentlessly and even selfishly.

But if there *is* a God, that would change *everything*. It would mean that we are not merely the product of biological chance but are the intentional creation of an unimaginably powerful Being. It would also infer ultimate accountability. It would mean that we are not the free agents we assumed ourselves to be, able to live as we please, but are 'guests' in Someone Else's universe. The possible existence of God brings with it the distinct possibility of our continued existence beyond physical death. If a metaphysical Supreme Being does exist, it opens the possibility of ongoing metaphysical existence for all of us.

Every other question that we might answer in this life – what career we will pursue and who we will marry – have limited consequences. At most, they might only effect 60 or 70 years. But the 'God question' has potential *eternal* consequences.

This is why C.S. Lewis, the great philosopher and theologian of the 20th century, commented:

> *"Christianity, if false, is of no importance, and if true, is of infinite importance. The only thing it cannot be is moderately important."*

I urge you to give serious attention to this question. The subject matter of this book is not merely idle speculation regarding a matter of intellectual curiosity: it is potentially a matter of life and death. It is the most important question in all of life.

It is my hope that as you read this book, you will be able to lay aside whatever natural antipathy and scepticism you may have and examine the evidence for God's existence with an open heart and mind.

2

REMOVING THE ELEPHANT

Before we even begin to examine the seven key areas of evidence for God's existence, we need to deal with the elephant in the room. There is a particular worldview that has become so entrenched within society that it is now regarded as an unassailable fact. It is a worldview that proposes a purely natural, mechanistic origin for life and, by inference, for the universe itself; a worldview that seems to preclude the necessity for a supernatural Creator and proposes that everything we see and experience can be explained by natural causes.

I am speaking, of course, about the theory of evolution.

Evolutionary theory has now embedded itself so deeply within society and within our collective conscience that it is a major stumbling block for many people in considering the possibility of God's existence. In fact, many people are so thoroughly indoctrinated with the theory of evolution that they are no longer willing to even entertain the possibility of God's existence. After all, evolution has disproved God hasn't it?

So, before we begin to examine the evidence for God's existence, I need to remove the elephant. I have previously written a whole book outlining the serious scientific flaws in the theory of evolution and detailing the growing tide of respected scientists from all over the world who have abandoned the theory of evolution entirely. The book is called *No More Monkey Business: Evolution in Crisis,* if you are interested in pursuing this topic in detail.

To help you consider the evidence for God's existence with an open mind, I firstly need to disavow you of the false assumption that evolution has somehow already 'disproved God'. I must remove the elephant from the room.

The first and most important thing to point out is that the theory of evolution is just that; a theory. It is not a fact. This is an extremely important distinction. In science, ultimately almost all "knowledge" remains within the realm of theory rather than fact. Even something as well-established as the theory of gravity, with its precise formula, cannot be said to have been proven in an ultimate sense, as we may one day find circumstances in either the macro or micro worlds (such as the sub-atomic quantum realm) where gravity behaves very differently. The formula for gravitational attraction may one day need to be amended to incorporate further variables. Thus, while we may colloquially describe gravity as "fact", a description of its precise nature and operation remains theoretical. Ultimately, all scientific theories remain permanently subject to either verification, amendment or disproof, through the examination of ongoing evidence.

Secondly, far from being substantiated by a vast body of evidence, the theory of evolution is rapidly losing traction and respectability within the scientific community due to the growing body of evidence that contradicts its central premises. Discoveries in the fields of genetics, microbiology, palaeontology, chemistry and other fields in recent decades have revealed the impossibility of Darwinian evolution to account for the complexity of life as we now understand it. Charles Darwin's naïve theory of the origin of species, proposed in 1859, is increasingly contradicted by current scientific evidence, with the result that the theory of evolution is now regarded as scientifically untenable by a large and growing body of respected scientists. Of course, this is not the impression given in popular media and nature documentaries, which continue to pump out evolutionist propaganda. But at the scientific coalface, where these matters are being studied at depth, there is growing discontent and disbelief in Darwin's theory.

Some of the evidence that now directly contradicts the theory of evolution includes:

- The failure of the fossil record to show gradual development of increasingly complex species.
- The impossibility of DNA forming through natural processes
- The impossibility of genetic mutations creating viable new biological features
- The now-acknowledged understanding that micro-evolution (small variations within a species) does *not* offer a viable means for macro-evolution (the formation of new species with *new* physical characteristics).
- The impossibility of living cells forming from non-living matter.

In response to these and many other recent scientific evidences, a growing number of respected scientists are voicing serious concerns regarding the theory, and some are abandoning it altogether. Conscientious objectors to evolution have been growing in number and becoming more vocal for over 50 years:

- In 1967, in response to the growing understanding of genetics, a scientific symposium was held at the Wistar Institute, a biomedical science research centre in Philadelphia, where the plausibility of evolution was challenged because of the now recognised impossibility of new physical features being created by random genetic adaptation.[1]
- In 1973, renowned palaeontologist, Dr Barbara J. Stahl, drew world attention to the failure of the fossil record to provide any unequivocal evidence for evolution, in her acclaimed book, *Vertebrate History; Problems in Evolution*.[2]
- In November 1980, at the Natural History Museum in Chicago, a large number of the world's leading geneticists and other scientists held a seminar to consider the issue of whether the small changes within a species, sometimes referred to as "micro-evolution", can lead to the big changes necessary for Darwinian evolution ("macro-evolution" - one species changing into a *brand-new* species, with new physical features). The findings of the conference were reported in the next issue of "Science" magazine, which stated:

"The central question of the Chicago conference was whether the mechanisms underlying micro-evolution can be extrapolated to explain the supposed phenomena of macro-evolution. At the risk of doing violence to the opinions of some of the scientists at the meeting, the answer was a clear 'No'."[3]

In other words, in 1980, a conference of the world's leading geneticists concluded that evolution is not genetically possible!

Removing The Elephant | 17

- In 1985, molecular biologist, Dr Michael Denton, in his landmark book, *Evolution: A Theory in Crisis*, outlined the newly emerged understanding of the irreducible complexity of the single cell; that even the simplest cell is a complex bio-factory of hundreds of interdependent molecular components, all of which need to be simultaneously existent and functional for the first cell to be alive. The theory of evolution does not have a viable explanation as to how these complex components could have sprung into existence simultaneously.[4]
- In the 1990s, the huge amounts of information in DNA within cells (which we will examine in a subsequent chapter) became a topic of great discussion. How could the 3.2 billion pieces of encoded information within *every* DNA strand in almost *every* cell of the human body have come into existence through natural processes? Where did this information come from? Evolution simply cannot account for this. Physicist, Dr Lee Spetner, highlighted these serious problems in his book, *Not By Chance: Shattering The Modern Theory of Evolution*, in 1997.[5] This extremely problematic issue has had evolutionists scratching their collective heads ever since. For example, Professor Werner Gitt, of the German Institute of Physics, published a critical book in 2006, entitled, *In The Beginning Was Information*.[6] This book outlined the complete inability of random evolutionary processes to produce the vast quantity of complex information within DNA. The existence of DNA is a huge problem for the theory of evolution!
- In 2002, in response to the mounting scientific evidence contradicting the theory of evolution, many of the world's leading scientists began to call for a symposium to determine its ongoing validity. As a

response, in that same year, an international organisation of scientists was formed, called CESHE (Cercle d'Etudes Scientifique et Historique), headquartered in France. After a period of intense scrutiny and rigorous evaluation of all the evidence, this was their conclusion:

"The theory of evolution is not supported by science. Many scientists have accepted the theory because they assume it to be an established scientific fact. Those scientists who have investigated it, however, find that evolution is a belief, not a science."[7]

- In July 2008, as the scientific integrity of the theory of evolution continued to unravel, a conference of the world's leading evolutionary scientists was held in Altenberg, Austria. The purpose of the conference was to discuss the growing realisation that if natural selection (the slight variations *within* a species - which is an observable and undeniable process) cannot produce *new* species with completely *new* physical features, then Darwin's theory is dead. The conference could not come up with a viable explanation of how natural selection could achieve this, given our current knowledge of genetics. After the conference, Dr Jerry Fodor, of Rutgers University, is quoted as saying, *"Basically I don't think anybody knows how evolution works."*[8]

These examples represent the tip of the iceberg regarding the growing dissatisfaction with the theory of evolution among the scientific community. Evolution is *far* from proven and an increasing number of respected scientists are daring to voice their disbelief.

In 2001, Australian biologist, Dr John Ashton, published the landmark book, *In Six Days; Why Fifty Scientists Choose To Believe in Creation*.[9] His book contains fifty chapters, each written by a different Ph.D. scientist. Each of them provides extensive scientific arguments for their view that the theory of evolution is no longer tenable, and they explain the growing evidence supporting the creation narrative. The contributing scientists are highly regarded internationally and come from a wide range of fields including biology, chemistry, biochemistry, genetics, physics, zoology, astronomy, meteorology, engineering and botany.

In 2012, Dr John Ashton published the book, *Evolution Impossible: 12 Reasons Why Evolution Cannot Explain the Origin of Life on Earth*.[10] The book provides a comprehensive discussion of the profound flaws that have become apparent in the theory of evolution in recent years.

In 2014, respected geneticist, Dr Casey Luskin, contributed a chapter to the book, *More Than Myth*, entitled *The Top Ten Scientific Problems with Biological and Chemical Evolution*.[11] The chapter received world-wide attention, prompting many other scientists to voice similar concerns.

In his book, *God, Science and Evolution*, E.H. Andrews, Professor Emeritus, University of London, wrote:

> "Speaking as a scientist, I believe that in another 20 years the theory of evolution will have been totally discredited, purely on scientific grounds. The enormous gaps in the theory are beginning to emerge – not, of course, in the popular versions of evolution, but in the findings of scientists who are studying these matters at depth."[12]

While popular media continue to give the impression of

consensus within the scientific community regarding the theory of evolution, the reality is that there is strong and growing disagreement. While the impression is given of complete confidence, the reality is an increasing atmosphere of doubt and uncertainty. Instead of a vast body of incontestable supporting evidence, the truth is that the evidence-base for evolution is dwindling rapidly, to the point where the theory is in serious crisis.

Prof. E.H. Andrews states,

> *"The popular impression is given that evolution is scientifically proven. This view is terribly biased and ignores the yawning chasms in the theory which make it unacceptable to me as a scientist."*[13]

EVOLUTION AND GOD

Having said all this, there are some people who continue to believe in evolutionary theory while also maintaining a belief in God. This is the concept of 'theistic evolution': the belief that God used the evolutionary process to create life. These people see no conflict between their Christian faith, and the process of evolution. I am not of this opinion, but there are plenty of intelligent people who hold this position. A survey of American churches in 2007 revealed that 51% of Protestant respondents and 58% of those who identified as Catholic believe that God used evolution to create life[14].

The reason some Christians have no problem believing in the theory of evolution is that it is not essentially a theory about origins, but about *processes*. It proposes a possible explanation about the processes by which the variety of species that we see today may have developed. But it has nothing to say about the origin of the universe itself. The theory of evolution assumes that the earth, and the whole universe, was already existent when organic life began, and it proposes no explanation for its origin. Furthermore, it fails to explain how a lifeless, inorganic universe could have produced biological life. For the answers to these key questions, Christians turn to the Bible and to their faith in a transcendent Creator God.

THE ELEPHANT HAS LEFT THE ROOM

You may still be unconvinced about the scientific flaws in the theory of evolution. You may still be a 'believer' in Charles Darwin's theory. I accept that a short chapter such as this may not erase a lifetime of evolutionary teaching. In that case, I recommend that you do some more research. I encourage you to continue to examine the current evidence for yourself. Here are some helpful resources:

- *No More Monkey Business: Evolution in Crisis*, Kevin Simington (Available from SmartFaith.net, Amazon and all online retailers)
- *Evolution Impossible*, John Ashton (Available from all online retailers)

But whether you continue to believe in evolution or not, I hope I have demonstrated to you that the theory of evolution DOES NOT preclude belief in God. Many people believe in both evolution and God (although I do not). Far from disproving the existence of God, the theory of evolution actually ***demands*** the existence of God, because our current scientific understanding reveals how utterly incapable evolutionary processes by themselves are for producing the complexity of life that we see today.

Evolutionary theory actually ends up asking more questions than it can answer.

The theory of evolution is not the 'God-slayer' that most people presume it to be. In fact, it is a toothless dinosaur that is mortally wounded and which has absolutely nothing to say about the existence of God at all. So don't let your belief in evolution stop you investigating the evidence for God's existence.

The elephant has left the room.

3

EVIDENCE 1: THE ORIGIN OF THE UNIVERSE

Without doubt, the most profound piece of evidence for the existence of God is, at the same time, science's greatest mystery: how could the physical universe have come into existence in the beginning? If, as science now proposes, the universe began with a 'Big Bang', where did the stuff that went 'bang' come from? And who or what made it go 'bang'?

Atheism offers no viable explanation as to how the physical universe came into existence. The theory of evolution starts with a *pre-existing* universe. It proposes a theory of how life supposedly progressed from simple to complex organisms, but it offers NO theories as to how the universe got here in the first place. In fact, at the time of Charles Darwin, who was among the first to propose the theory of evolution, it was simply assumed that the universe had always been here; that it had existed eternally.

We now know this is not the case at all. A succession of remarkable discoveries in the second half of last century demonstrated that the universe has NOT always existed. It is now almost universally accepted among cosmologists and astrophysicists that the universe actually had a beginning; that at some point in the distant past, there was nothing, and then, a 'moment' later, there was a universe!

This raises the obvious question: how did the universe come into existence? Who or what created it? For, surely, it can't have created itself!

THE EVIDENCE OF COSMOLOGY

Cosmology is the study of the cosmos - the universe, with its untold billions of galaxies and stars. For the Christian, who looks up in wonder into the night sky, the sheer scale and majesty of the cosmos has always been regarded as a strong evidence for the existence of an almighty Creator. The Songwriter in the Bible declared:

> *"The heavens declare the glory of God; the skies proclaim the work of his hands" (Psalm 19:1)*

In contrast, the response of the atheist scientist has been to simply claim that the universe has always been there. For centuries the Christian has asked the atheist, *"Who made the universe?"*, to which the unbeliever has replied, *"No one! It has simply always existed!"* Outspoken atheist, Bertrand Russell (1872 - 1970) famously stated;

> *"The universe is just there, and that's all!"*[1]

This belief in the eternal existence of the universe was the prevailing view of scientists and natural philosophers from the time of Aristotle (350 B.C.) until the beginning of the 20th century. This completely baseless presupposition was extremely convenient for atheists, because it effectively

Evidence 1: The Origin Of The Universe | 25

removed the need to believe in any sort of creator. A series of stunning discoveries in the modern era, however, completely overturned this view of the universe.

After the publication of Albert Einstein's Theory of General Relativity in 1915, Dr Willem de Sitter published extrapolations of Einstein's theory in 1917, predicting that, if Einstein's theory is correct, the universe should be expanding outwards and, conversely, must have had a finite beginning.[2] This idea of a universe that had a beginning was contested vigorously by the scientific community at large, including Einstein himself, who wrote to Dr de Sitter, stating, "*This circumstance irritates me*" and in a second letter he stated, "*To admit such possibilities seems senseless.*"[3] The scientific community sided with Einstein, and largely ignored Dr de Sitter's observations.

In 1927, cosmologists Dr Alexander Friedmann, Fr. Georges Lemaitre, Dr Howard P. Robertson and Dr Geoffrey Walker published more detailed extrapolations of Einstein's theory, (referred to as the Friedmann–Lemaître–Robertson–Walker metric)[4], confirming Dr de Sitter's findings that the universe must be expanding and lending further weight to the argument that it must have had a beginning.

In 1929, Dr Edwin Hubble (1889 - 1953), confirmed the expansion of the universe by observing the Doppler redshift of the light from observable galaxies, indicating their recession from earth.[5] In other words, he confirmed from physical observation that the galaxies in the visible cosmos are expanding outwards.

The discovery of cosmic microwave background radiation (CMBR) in 1964 by Drs. Arno Penzias and Robert Wilson[6] (which earned them a Nobel Prize in Physics in 1978) provided the scientific community at that time with what seemed conclusive evidence of residual microwave radiation left over from a "big bang" at the beginning of space-time. Thus, the Big Bang theory of the origin of the universe was formed. This concept that the universe had a beginning, represented a complete reversal of a belief in its *eternal* existence which had been passionately held for over 2,000 years. Dr Stephen Hawking (1942 - 2018), in a lecture published on his website, commented on this reversal:

> "All the evidence seems to indicate, that the universe has not existed forever, but that it had a beginning. This is probably the most remarkable discovery of modern cosmology."[7]

BIG BANG IN TROUBLE

In recent decades, some serious doubts have been cast upon the validity of a "big bang" as an explanation of the universe's creation. One of the major problems is explaining how such a big bang could account for the formation of galaxies and clusters of galaxies, which we observe throughout the universe. If a big bang created the universe, it should have resulted in a fairly uniform dispersal of matter throughout the universe, resulting in a sparsely spread scattering of matter. But this is not what we see. Instead, we find huge clusters of galaxies, densely populated by stars, separated by vast distances of empty space. No computer models of a supposed big bang can show how these clusters could possibly have formed. Dr James Trefil, Professor of Physics at George Mason University, Virginia, comments:

> "There shouldn't be galaxies out there at all, and even if there are galaxies, they shouldn't be grouped together the way they are.' He later continues: 'The problem of explaining the existence of galaxies has proved to be one of the thorniest in cosmology. By all rights, they just shouldn't be there, yet there they sit. It's hard to

convey the depth of the frustration that this simple fact induces among scientists."[8]

Because of these and other observational anomalies that seem to contradict the Big Bang theory of the universe's origin, a growing number of scientists are calling for the theory's official demise. For example, in 1993, cosmologist Dr Halton C. Arp, of the Mount Wilson Observatory, Pasadena, USA, wrote:

> *"In my opinion the observations speak a different language; they call for a different view of the universe. I believe that the big bang theory should be replaced, because it is no longer a valid theory."*[9]

The expansion of the universe has also been challenged recently by some observational data suggesting alternate explanations for redshift in galaxies and for cosmic microwave background radiation (CMBR).[10]

These and other recent developments have led a growing number of respected scientists to reject the notion of a big bang. In 2004, an "Open Letter To The Scientific Community" by 33 leading scientists was published on the internet,[11] and republished in *"New Scientist"*.[12] A subsequent article was published on www.rense.com, entitled *"Big Bang Theory Busted by 33 Top Scientists"*.[13]

Modern cosmology is, therefore, a long way from consensus on these issues. Despite the divergence of opinions, however, there is a very clear groundswell of scientists willing to concede the existence of a creative force or forces instrumental in the universe's formation. Even those who continue to hold to the big bang theory as the cause of the universe's beginning cannot adequately explain the observable complexity of the universe. Something else had to be at work, shaping the universe at the beginning of time.

THE UNIVERSE HAD A BEGINNING

Despite controversy and growing scientific scepticism regarding the big bang as an *explanation* for the universe's beginning, most scientists agree that the evidence of modern physics and cosmology points inexorably to the conclusion that the universe began *somehow*. The process of that beginning is unclear, but its fact is now almost universally accepted. This is further confirmed by several other very powerful theoretical arguments.

The Second Law of Thermodynamics

Newton's second law of thermodynamics indicates that in any closed system, the total amount of available energy diminishes over time. This is referred to as entropy, and, when applied to cosmology, it simply means that the universe is gradually winding down. It also means that over time the universe is getting progressively colder. Our observation of the nature and function of stars confirms this. Because stars are finite bodies with finite mass and energy, it is not possible for them to burn continuously for eternity. Given sufficient time, each star will eventually burn itself out of existence. This is a simple logical extension of Newton's theory. If the universe has existed for eternity past, by this argument it would have burnt itself out by now!

The Philosophical and Mathematical Impossibility of an Eternal Universe

Al-Ghazali, a Persian philosopher in the Middle Ages, proposed this simple philosophical proof that the universe cannot have existed forever. He stated:

"If there were an infinite number of events in the past, we would never have arrived at the present"[14]

This simple, yet profound philosophical argument cannot be refuted without dismissing the laws of logic. Think about it. If time is a corridor and you started in the present and went back in time towards infinity past, you would never reach infinity past. You can't reach infinity past from the present. You will never get there! In the same way, you can't reach the present from infinity past. It's impossible! Al-Ghazali's theorem proves that the universe cannot have existed forever.

Similarly, Al-Ghazali also proposed that:

"An infinite number of things cannot exist."[15]

This is because infinity cannot logically exist in a physical universe. David Hilbert was a German Mathematician who came up with a brilliant way of explaining this concept. He asks us to imagine a hotel with an infinite number of rooms, each occupied by a guest. Therefore, the number of guests is infinity. Suppose one more guest arrives and wants a room. The hotel manager asks every guest to move to the next room number. The guest in room 1 moves to room 2. The guest in room 2 moves to room 3. The guest in room 5,000,000,000 moves to room 5,000,000,001 and so on. Now there is a vacant room in Room 1, and the new guest is accommodated. But the number of guests still numbers infinity.

Thus:

$$\text{Infinity} + 1 = \text{Infinity}$$

This works for the addition of any finite number. It also works for subtraction.

But suppose an infinite number of new guests arrive, each requiring a room. No problem! Every existing guest is simply asked to move to a room number twice their existing number. Thus, Room 1 moves to room 2. Room 2 moves to room 4. Room 3 moves to room 6, and so on. In this way the pre-existing infinity of guests are now all in even numbered rooms, leaving an infinity of odd numbered rooms vacant. The infinity of *new* guests can now move into the odd numbered rooms.

But how many guests are there now? Still infinity! Thus:

$$\text{Infinity} + \text{infinity} = \text{infinity}$$

But suppose everyone in the even rooms now leave? An infinite number of guests were in the hotel. An infinite number of guests now leave (even numbers) but an infinite number of guests still remain (odd numbers). Thus,

$$\text{Infinity} - \text{infinity} = \text{infinity}$$

What on earth has all this got to do with the origin of the universe? Quite simply, it shows how absurd infinity is as an actual number in a physical universe. Infinity exists only as an abstract concept; it cannot exist as an actual number in a physical universe. Therefore, there cannot have been an infinite number of past events and past days, hours or seconds. Therefore, the universe must have had a beginning!

The evidence of cosmology, philosophy and mathematics all points to a clear, logical conclusion: The universe has not existed forever. It had a beginning. As Dr Stephen Hawking stated on his website:

> *"We have made tremendous progress in cosmology in the last hundred years ... which has shattered the old picture of an ever-*

existing and ever-lasting universe. ... This is a profound change in our picture of the universe and of reality itself."

THE NECESSITY OF A SUPERNATURAL CAUSE

This brings us to the logical conclusion; something *beyond* the physical universe had to have created it.

Most cosmologists are now reaching the astonishing conclusion that at some time in the distant past there was nothing at all. No stars, no planets, no asteroids, no gases, no chemicals, no elements, no physical matter at all. Nothing. And then, a 'moment' later, there was a universe! Stephen Hawking was correct in saying that this is *"a profound change in our picture of the universe and of reality itself."* Because if the universe had a beginning, it raises the question: Who put it there? How can *nothing* become *something* unless *someone* or *something* beyond the universe creates it? The cosmological evidence for the universe having a beginning cries out for a supernatural explanation, because only something *outside* of nature, something 'super-natural', can possibly create nature. Nature cannot create itself out of nothing!

THE KALAM COSMOLOGICAL ARGUMENT

The Kalam cosmological argument deals with the issue of ultimate cause. It is a philosophical argument, originating in the Middle Ages, and championed in recent years by apologist and theologian, Dr William Lane Craig. The original Kalam argument is as follows:

1. Whatever begins to exist has a cause.

2. The universe began to exist.
3. Therefore, the universe has a cause.

William Lane Craig added to the argument:

1. That cause must be timeless, immaterial, self-existent and powerful.
2. The most reasonable cause is God.

This is a logical argument that, in my opinion, is impossible to refute. The universe can't have created itself, therefore the cause must lie outside of the universe! Of course, there will always remain a core of obdurate atheists who refuse to concede even this clear chain of logic. For example, Dr Quentin Smith, Professor Emeritus of Philosophy at Western Michigan University, in a debate with William Lane Craig, stated:

> "The universe came from nothing, by nothing for nothing!"[16]

This, of course is a ludicrous proposition, and it illustrates that some atheists are prepared to believe in the impossible, rather than the supernatural.

A growing number of scientists, however, are concluding that there was something beyond nature that was fundamentally at work in the creation of the universe. In April 2016, Dr Dan Reynolds wrote:

> "All the observable evidence we have about the universe implies it had a beginning ... Logically, the universe did not and could not create itself. If the universe (nature) could/did not create itself and it had a beginning, then only something or someone outside of nature can account for the universe's existence. Genesis 1:1 offers a credible

Evidence 1: The Origin Of The Universe | 33

explanation: In the beginning God created the heaven and the earth."[17]

Dr Robert Jastrow, astronomer, physicist and founder of NASA's Goddard Institute of Space Studies, stated,

> "Astronomers now find they have painted themselves into a corner because they have proven, by their own methods, that the world began abruptly in an act of creation to which you can trace the seeds of every star, every planet, every living thing in this cosmos and on the earth. And they have found that all this happened as a product of forces they cannot hope to discover. That there are what I or anyone would call supernatural forces at work is now, I think, a scientifically proven fact."[18]

Similarly, Dr James Clerk Maxwell, physicist and mathematician, who is credited with formulating classical electromagnetic theory and whose contributions to science are considered to be of the same magnitude to those of Einstein and Newton, stated:

> "Science is incompetent to reason upon the creation of matter itself out of nothing. We have reached the utmost limit of our thinking faculties when we have admitted that because matter cannot be eternal and self-existent it must have been created."[19]

Commenting on the growing number of scientists who now concede that the universe must have had a supernatural cause, astrophysicist, Dr Hugh Ross, Director Emeritus of Observations at Royal Astronomical Society, Vancouver, states;

> "Astronomers who do not draw theistic or deistic conclusions are becoming rare, and even the few dissenters hint that the tide is against them. Geoffrey Burbidge, of the University of California at

San Diego, complains that his fellow astronomers are rushing off to join 'the First Church of Christ of the Big Bang.'"[20]

A growing tide of scientists at the top of their fields, when confronted with the mounting cosmological evidence, are conceding that the only logical explanation for the origin of the universe, is that there must have been a supernatural cause. Nature cannot create itself; therefore the cause had to have been something outside of nature - a transcendent *"supernatural"* cause. This does not mean that all the scientists in the world are suddenly becoming Bible believers. There is a big step from believing in a supernatural cause of some kind, to believing in the God of the Bible. But for the first time in a long time, a growing chorus of voices within the scientific community has conceded the very real possibility of the existence of supernatural forces that lie beyond the realm of scientific study.

Dr Hugh Ross summarises;

> "All the data accumulated in the twentieth and twenty-first centuries tell us that a transcendent Creator must exist. For all the matter, energy, nine space dimensions, and even time, each suddenly and simultaneously came into being from some source beyond itself. Likewise, it is valid to refer to the Creator as transcendent, for the act of causing these effects must take place outside or independent of them."[21]

Of course, this is not the impression we get from the media. Humanists seem to have a stranglehold on the popular press. Scientists with aggressive atheistic agendas dominate the airwaves, and their well-financed documentaries continue to pump out the message that science has a natural explanation for everything. The impression is given that the 'new god of science' has all the answers sewn up. But scientists who are at the cutting edge of their fields know that this is not so. Those who are studying these things in depth are coming face-to-face with undeniable evidence of a transcendent, supernatural reality that underpins our entire universe. The tide is turning, and it's not just a trickle. The previously quoted complaint of Dr Geoffrey Burbidge, of the University of California at San Diego, that his fellow astronomers are *"rushing off to join the First Church of Christ of the Big Bang"*[22] indicates how widespread this new supernatural awareness is within the scientific community.

Without a doubt, the science of cosmology provides extremely convincing evidence for the existence of a supernatural creator-God. The extraordinary claim by Richard Dawkins, that *"there is not a tiny shred of evidence for the existence of any kind of god"*[23], must surely arise from a wilful determination to ignore the considerable cosmological evidence that has arisen in recent years, and portrays his intransigent unwillingness to even consider the existence of anything beyond the realms of empirical science.

THE DESPERATE SEARCH FOR AN ALTERNATE EXPLANATION

As the evidence of cosmology increasingly points to a supernatural cause, die-hard atheists are becoming ever more desperate in their search for an alternate explanation for the origin of the universe that does not involve a personal God. These alternate explanations include:

- **Bubble universes:**[24] Also known as "eternal inflation theory". An endless series of new bubble universes, expanding and breaking off from existing universes,

like giant boils. (But where did the first universes come from?)
- **An oscillating or cyclical universe:**[25] The idea that our universe is endlessly expanding then contracting back to a single point again, then exploding in a big bang again, on and on forever. (But logic and the second law of thermodynamics have both proved that a physical universe cannot be eternal!)
- **Baby universes:**[26] One of Stephen Hawking's desperate postulations theorised that black holes are the umbilical cords that connect baby universes with the mother universes that gave birth to them. (But where did the mother universes come from?)
- **Aliens created our universe:**[27] Proposed by Richard Dawkins and others. (Oh dear! Really? In that case, who created the aliens?)
- **Our future selves created the universe:**[28] Our future selves eventually developed time travel and travelled back to the distant past and created the physical universe for our past selves to evolve into. (I'm not making this up! This has been suggested in scientific circles!)
- **Simulation theory:**[29] Our universe is simply a giant computer simulation designed by aliens or, once again, our super-advanced future selves, and we are all simply plugged into the simulation. This theory was proposed by Dr Nick Bostrom, Professor of Physics at Oxford University in 2003.[30] ("The Matrix" is apparently true!)
- **Something from nothing:** In another of Stephen Hawking's desperate attempts to avoid the existence of a creator-God, he postulated, *"Because there is a law such as gravity, the universe can and will create itself from nothing."*[31] (This effectively throws all the laws of

science and the accepted laws of cause and effect out the window!)
- **We don't know:** Those less inclined to scientific fairy stories, but still equally dismissive of the supernatural, simply state that *nobody* knows how the universe was initially formed. An article on the website, *American Scientist*, commented: *"The question of how matter came into existence in the formation of the universe still awaits a satisfactory answer."*[32] (At least this is honest!)

Christopher J. Isham, Britain's leading quantum cosmologist, and an astrophysicist at Imperial College of London, recently wrote:

> *"The idea that the Big Bang supports theism is greeted with obvious unease by atheist physicists. At times this has led to <u>wild scientific theories</u> being advanced with a tenacity which so exceeds their intrinsic worth that one can only suspect the operation of psychological forces lying very much deeper than the usual academic desire of a theorist to support his or her theory."*[33] (Emphasis mine).

Dr Isham is absolutely correct. These wild and bizarre theories indicate a desperate determination to avoid belief in a supernatural Creator, and a corresponding willingness to embrace anything, no matter how fanciful, in order to do so. As a Christian, this does not surprise me in the least. As the Apostle Paul wrote:

> *"The god of this age has blinded the minds of unbelievers, so that they cannot see the light of the gospel that displays the glory of Christ, who is the image of God."* (2 Cor 4:4)

More recently, in a variation of the 'something from nothing' proposition, some cosmologists and physicists have theorised that the physical universe may have originated from fluctuations in some kind of pre-existing quantum energy field. Drs. Jim Hartle, Stephen Hawking and Alex Vilenkin have all been proponents of various iterations of this theory, which effectively redefines the 'nothing' from which the physical universe sprang as a chaotic space-time foam with fantastically high energy density. Of course, this begs the question, "Where did THAT supposed quantum energy field come from?". Theories such as this simply move the question of ultimate cause back an additional step. One must still ask, *"How did ANYTHING come into existence?"* Because, as we have already seen, the mathematical impossibility of an infinite series of events or increments of time asserts that there MUST have been a beginning.

For this reason, the science philosopher Dr John Leslie explains that NONE of today's exotic speculations regarding the origin of the physical universe from fluctuating quantum energy fields precludes the possibility of God's existence. Even if true, they would simply explain the means by which God might have created the universe.[34]

Soon after his book, *'A Brief History of Time'*, was published, Stephen Hawking was interviewed and was asked if his latest theories regarding quantum fluctuations giving rise to physical matter precluded the existence of a Creator. Hawking admitted that such theories did *not* preclude God's possible existence. In another of his writings, Hawking also admits that even if we could understand the physical laws that might have caused the universe to have been created, those laws would not necessarily be a disproof of God's existence, but could simply show that God did not choose *"to set the universe going in some arbitrary way that we couldn't understand. It says nothing about whether or not God exists – just that he is not arbitrary."*[35]

It must be stressed, however, that these latest theories about fluctuations in quantum energy fields creating physical matter are wildly speculative theories arising more from imaginative dreaming rather than from any empirical scientific evidence. The puzzle as to how the physical universe could have come into existence remains science's greatest mystery. At the same time, it is also arguably the most profound evidence for the existence of a supernatural Creator.

Dr Francis Collins, in his book, *The Language of God: A Scientist Presents Evidence for Belief*, states:

> *"We have this very solid conclusion that the universe had an origin ... The universe began with an unimaginably bright flash of energy from an infinitesimally small point. That implies that before that, there was nothing. I can't imagine how nature, in this case the universe, could have created itself. And the very fact that the universe had a beginning, implies that someone was able to begin it. And, it seems to me, that had to be outside of nature."*[36]

To summarise this first area of evidence:

1. The universe had a beginning.

2. The universe can't have created itself.

3. The cause of the universe must, by definition, exist beyond nature; it must be *supernatural*.

4. The best and most logical explanation is the existence of a supernatural Creator God.

The logic of this argument is difficult to refute. In fact, given what we now know regarding the fact of the universe's beginning, it takes MUCH more faith to hold that there is no God than to believe in one. In rejecting the existence of God, you are left with the impossible scenario of the universe creating itself – a scenario which has resulted in a series of ridiculous and desperate theories by atheists who are determined to cling to their unbelief despite any and all evidence to the contrary.

Of course, many with open hearts and minds have been able to perceive God's hand in the creation of the universe. Robert Boyle (1627 – 1691), known as the founder of modern chemistry, once stated,

> "God is the author of the universe, and the free establisher of the laws of motion."[37]

Similarly, Max Planck (1858 - 1947), the theoretical physicist who won the Nobel Prize in Physics in 1918 and was the originator of quantum theory and the developer of "Planck's Constant" (an extremely important constant in cosmology), also came to the inescapable conclusion that God was the originator of the universe:

> "God is the beginning and end of all considerations: He is the foundation."[38]

What about you? Are you open to considering this compelling evidence? As Dr Antony Flew said, *"We must follow the evidence wherever it leads."* And in this case, the evidence leads us directly to the existence of an all-powerful, supernatural Creator God who called the universe into existence.

4

EVIDENCE 2: THE ORIGIN OF LIFE

One of the most powerful evidences for the existence of God is the mystery surrounding the origin of the single cell. Evolutionary theory supposes that all biological life arose from single-celled life in the distant past. Arguably, the theory's greatest stumbling block has always been to explain how the first living cells developed; how inanimate chemicals gave rise to living cells through natural, chance processes. Because if evolution is true and there is no God, then that is what must have happened. Proving HOW this could have happened has become the holy grail of evolutionary theory.

Abiogenesis is the term used to describe the supposed original evolution of cellular life from inorganic, inanimate matter. Earlier last century the concept of lightning striking a primordial soup of chemicals was thought to be sufficient for abiogenesis to have occurred – for the first living cells to have been created from non-living chemicals – but as time has passed and the fields of genetics and microbiology have progressed, the absurdity of that proposition has become increasingly apparent. The single cell is now understood to be unimaginably more complex than the scientists of the early 20th century envisaged.

In particular, there are three aspects of the living cell's complexity that continue to defy any naturalistic explanation regarding their origin and which point us to the existence of an all-powerful Creator God:

- The irreducible complexity of the cell's physical structures

- The irreducible complexity of the cell's molecular 'machines'

- The impossibility of proteins forming by chance chemical processes

Each of these will now be examined in detail.

THE IRREDUCIBLE COMPLEXITY OF THE CELL'S PHYSICAL STRUCTURES

Until the middle of last century, the living cell was regarded as a simple blob of protoplasm surrounded by a membrane. Thus, it did not seem unreasonable to suppose that the first cells could have formed via random chemical processes, given sufficient time and the right conditions. Modern advances in the field of microbiology, however, have dramatically transformed our understanding of the living cell. Scientists now realise that it is unimaginably more complex than they once regarded it, rendering the concept of abiogenesis – the first cells forming via natural processes – extremely improbable. Indeed, as you will see, a growing number of scientists are now declaring abiogenesis to be impossible.

A major setback for proponents of abiogenesis has been ongoing discoveries regarding the irreducible complexity of the living cell. 'Irreducible complexity' is a term coined by

Evidence 2: The Origin Of Life | 43

biochemist Dr Michael Behe in his book, *Darwin's Black Box: The Biochemical Challenge to Evolution*[1], and has subsequently been embraced by many other biochemists. It refers to any complex system that cannot be broken down into simpler components without losing its functionality. In the case of the living cell, it is argued that the cell is comprised of a large number of components which are essential to its viability as a living cell. Michael Behe and others argue that these components could not have evolved by a succession of small, incremental modifications because without each of these components in place and fully functional from the very beginning, the supposed primordial cell would not have been alive and would not have been able to replicate. In essence, irreducible complexity posits that each essential element of the living cell had to be in place and fully functional from the very beginning; they had to be created simultaneously in the very first living cells.

Michael Behe refers to a mousetrap to illustrate his point.

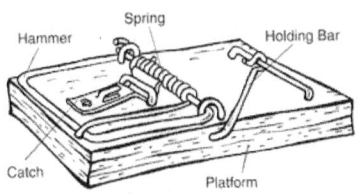

A simple spring-operated mousetrap has just five interdependent parts: a wooden platform, the spring, the hammer (which pins the mouse against the wooden base), the holding bar and a catch. Each of these components is absolutely essential for the mouse trap to function. Without any one of these, the mouse trap would not simply be less effective, it would not function at all. The mouse trap is irreducibly complex. Each of the components must exist and be in their correct positions simultaneously, in order for the mouse trap to function.

The problem for proponents of abiogenesis is that the irreducibly complex living cell is comprised of a lot more than just four essential components. Far from being a blob of shapeless protoplasm, a cell is an intricately designed entity comprised of many complex interdependent structural elements. Consider the following diagram of the structural components of a typical cell:

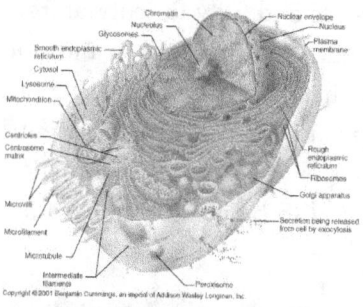

A typical cell has approximately 30 different structural elements, all playing a vital role in its viability. If the cell had no mitochondria for converting nutrients into energy molecules, it would die. If it had no cell wall, it would die. If it had no nucleus with its repository of DNA to provide the cell with its operating instructions, it would die. In fact, remove any of the structural elements from a functioning cell, and it will cease to be viable and, in almost all cases, will die without replicating. All of these structural components had to be in place from the very beginning; they all had to come into existence simultaneously in order for the first cell to exist as a living, reproducing biological entity. The physical structures of a cell could not have developed gradually and incrementally over time, because any supposed partially-formed primordial cells with partially-formed physical structures would not have been a living entity capable of replication and would, therefore, have been unable to pass on its half-formed structures for further development. This is now widely accepted among microbiologists. In this sense, the cell can be said to be structurally irreducibly complex.

As Dr Tas Walker states:

> "The whole structure of the living cell points to it being irreducibly complex."[2]

The problem facing scientists who are intent on providing an entirely naturalistic explanation for the origin of living cells, is that the incredibly complex physical structures within cells could not have come into existence through the random mixing of chemicals in a supposed primordial soup. Nor could such random chemical combinations explain the precisely engineered integration of these physical structures into a living, functioning entity. The irreducible complexity and design of the living cell is an extremely powerful argument for a Creator God.

THE IRREDUCIBLE COMPLEXITY OF THE CELL'S MOLECULAR MACHINES

But we have barely scratched the surface of the cell's irreducible complexity. These are merely the gross structural elements of a cell that all need to be in place for it to be a functioning entity. The development of electron microscopes in the latter part of last century allowed microbiologists to delve further into the cell at the molecular level, and what they discovered was truly astounding. Moving around within the physical structures of the cell is a vast army of living molecules operating like tiny robots, carrying out a variety of functions that keep the cell alive.

Simple cells are now understood to be extraordinarily sophisticated biological factories, teeming with hundreds of tiny molecular machines scurrying throughout the cell, carrying out hundreds of specialised tasks, all of which are essential for the cell to function effectively and remain viable. Some molecular machines transport nutrients from outside the cell, through the cell wall into the cell. Some convert energy stores into usable nutrients. Some carry nutrients throughout the cell. Some transport waste back to the cell wall. Some open up chemical holes in the cell wall to allow waste to be evacuated. Some repair other machines. Some build new machines. Some build the components of new cells. The living cell is literally swarming with teams of specialised molecular machines!

For example, there is the extraordinary assembly line of molecular machines within the nucleus of each cell, responsible for producing new proteins:

- Molecular machines to uncoil a strand of DNA ready for transcription
- RNA polymerase machines to transcribe a section of DNA into messenger RNA (mRNA)
- Molecular machines to cut and splice sections of the mRNA
- Molecular machines to transport the mRNA outside the nucleus into the cytoplasm of the cell
- A molecular machine known as ribosome which binds to the mRNA and reads the code in the mRNA to produce a chain made up of amino acids
- Transfer RNA molecular machines to link new amino acids to the mRNA chain in the correct order (thus forming a new protein),

- Molecular machines to then fold the new protein into its correct shape,
- Molecular machines to carry the new protein to the site in the cell where it is needed.

The image below is a computer-generated model of the inside of a cell, produced by Harvard University, showing the spherical nucleus of the cell surrounded by the intra-cellular structures and molecular "highways".[3] Travelling along these highways, hundreds of molecular machines can be seen moving throughout the cell, carrying out their vital functions.

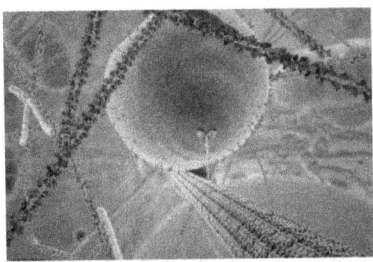

(Inside a living cell. Image: courtesy of Harvard University)

The cell's molecular machines are essential to the proper functioning and survival of the cell, and without all of them working together, the cell would die. In this sense, the humble cell can be said to be functionally irreducibly complex. Its viability – its survival and ability to replicate – is absolutely dependent upon the simultaneous co-existence of this vast army of interdependent machines. Furthermore, the cell could not have developed this army of molecular factory workers gradually and incrementally over time, because they are all interdependent; if any group of these machines was not in existence in the first cell, the cell would not have been alive and able to replicate.

But we have still not reached the full extent of the humble cell's irreducible complexity. Each of these molecular machines is, itself, irreducibly complex, comprised of precisely engineered individual proteins with very specific shapes and functions which form the integral components of the functioning machine. Some molecular machines even have parts that function as precisely engineered cogs and gears and rotors and propellers, which allow the machines to move and rotate in order to carry out their specific functions within the cell. Indeed, the term 'machine' is entirely appropriate, because these molecular machines, when viewed with electron microscopes, look like precisely engineered machines that have been manufactured in an engineering workshop!

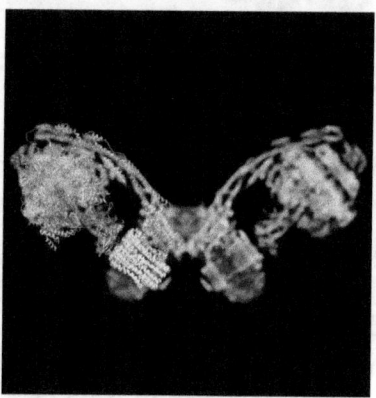

ATP Synthase Molecular Machine. (Image courtesy of sciencedirect.com)

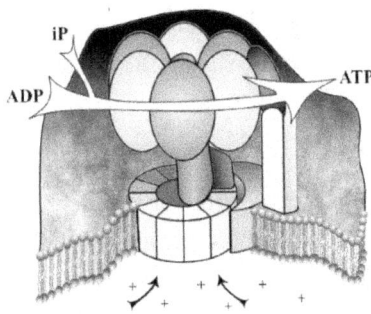

For example, the ATP synthase (pictured above, firstly as a photograph from an electron microscope and secondly as a diagrammatic image) is one of the more spectacular molecular machines that populate living cells. Respected biologist, Dr Alex Williams, explains:

> "It is a proton-powered motor that produces the universal energy molecule ATP (adenosine tri-phosphate). When the motor spins one way, it takes energy from digested food and converts it into the high-energy ATP, and when the motor spins the other way, it breaks down the ATP in such a way that its energy is available for use by other metabolic processes."[4]

These molecular machines within a living cell are irreducibly complex, because all of their component parts must be present for the machine to work. For example, the ATP synthase machine (above) requires all of its component parts to be fully present in order to function. Without the cogs at its base, or its central rotor shaft, or its motor mechanism, it would not function, and the cell would be without a source of energy to survive. All the parts of EVERY molecular machine had to be present from the very beginning and could not have developed via incremental successive steps, because the machines would simply not function if they were not complete. Evolutionists have been unable to propose a verifiable means by which any of these irreducibly complex molecular machines within a cell could have come into fully functional existence via natural means.

Thus, the 'humble' cell presents us with a cascading diorama of irreducible complexity. In order for the first cell to have formed via random natural processes, about thirty fully-formed gross structural elements all had to be in place and fully formed. Furthermore, for the cell to be actually 'alive' rather than merely a lump of tiny structural matter, the first cells had to come into existence with an army of fully-functioning molecular machines, all of which are essential to the cell's viability. And, finally, each of those machines needed to be formed with all of their precisely engineered protein parts fully formed and in the right place in order for them to be functional.

This cascading series of irreducible complexity poses huge problems for scientists seeking to explain how the living cell could have originated via chance chemical processes. Indeed, it defies all reasonable explanation how a fully-formed, fully-functioning, irreducibly complex cell could have popped into existence in a primordial soup via natural means. No amount of lightning strikes into a chemical soup could create such a highly structured living biological factory!

The previous claims of evolutionary scientists, that living cells must have developed via gradual, incremental steps, has been totally discredited by modern discoveries of the cell's irreducible complexity. Biochemist, Dr Jonathan Sarfati, in chapter 10 of his 2013 book, *Refuting Evolution 2*, writes:

> "*Living things have fantastically intricate features - at the anatomical, cellular and molecular level - that <u>could not function if they were any less complex</u> or sophisticated.*"[5] (Emphasis mine).

Dr Michael Behe's book, *Darwin's Black Box: The Biochemical Challenge to Evolution*[6] was a ground-breaking academic work arguing for the impossibility of evolution to create irreducibly complex systems. In it he explains the main argument against evolution:

> "By 'irreducibly complex' I mean a single system composed of several well-matched, interacting parts that contribute to the basic function, wherein the removal of any one of the parts causes the system to effectively cease functioning. An irreducibly complex system cannot be produced directly (that is, by continuously improving the initial function, which continues to work by the same mechanism) by slight, successive modifications of a precursor system, because any precursor to an irreducibly complex system that is missing a part is by definition non-functional. An irreducibly complex biological system, if there is such a thing, would be a powerful challenge to Darwinian evolution."[7]

Since its publication, *Darwin's Black Box* has elicited furious debate from evolutionists who claim to be able to refute his arguments, but their arguments are hypothetical, having no evidence to support them, and revolve around semantics, playing with the meaning of words. To date, no evolutionist has been able to offer a practical, observable, verifiable process whereby a prolonged series of random, successive, *non-functional* modifications can lead to a fully-functioning irreducibly complex system. In response to his critics, Dr Behe and other intelligent design advocates repeatedly point out that the gradual, successive, development of non-functioning components would be of no advantage to a cell until all the components are present; and without all the components present, the cell simply could not survive and replicate in order to pass on those partial modifications![8]

Dr Michael Behe is just one of many scientists today who are proclaiming the impossibility of the cell's origin via natural means. An article published on the Intelligent Design and Evolution Awareness Centre (IDEA) website, examines, among other examples of irreducible complexity, the cilia on the outside of single-celled organisms; hair-like filaments used to propel a cell through fluid. The article states:

> "*Evolution simply cannot produce complex structures in a single generation as would be required for the formation of irreducibly complex systems. To imagine that a chance set of mutations would produce all 200 proteins required for cilia function in a single generation stretches the imagination beyond the breaking point. And yet, producing one or a few of these proteins at a time, in standard Darwinian fashion, would convey no survival advantage because those few proteins would have no function - indeed, they would constitute a waste of energy for the cell to even produce. Darwin recognized this as a potent threat to his theory of evolution - the issue that could completely disprove his idea. So the question must be raised: Has Darwin's theory of evolution "absolutely broken down?" According to Michael Behe, the answer is a resounding 'yes'.*"[9]

THE IMPOSSIBILITY OF PROTEINS FORMING BY CHANCE CHEMICAL PROCESSES

In the previous section we examined the physical structures of the cell and then 'zoomed in' to examine the army of molecular machines that are in constant operation within those structures. Now we are going to increase the magnification even further and zoom down to examine the tiny building blocks of the cell's structures and molecular machines: proteins. Dr John

Ashton points out that a single cell has 2.4 million protein molecules made up of approximately 4,000 different types of proteins.[10] Furthermore, he points out that ongoing attempts to create even a single protein in laboratories with the latest technologies, have completely failed.[11]

Of course, the fall-back position of evolutionists is that, given enough time, random processes must eventually result in the right chemical combinations to bring about life. Dr Clifford D. Sirnack, in his book, *Trilobites, Dinosaurs and Man*, exemplifies this argument:

> *"If we had amino acids, we then would have proteins, and if we had proteins, we would be well along the road to life. Given trillions upon trillions of possibilities for chemical combinations, given a few million years for it all to happen, the components of life would have appeared. And once that had been accomplished, once the bricks and the stones and the lumber for the building of life were present, then all that would have been required were a few more million years for life to actually appear."*[12]

This type of argument, which persists to the current day, appears, on the surface, to have merit. After all, surely given enough time, all the necessary components of cellular life can form by chance.

However, this line of reasoning vastly underestimates the statistical improbability of even a single protein evolving by chance. In 1962, Dr Isaac Asimov (a professed atheist) calculated the probability of a simple insulin-like protein evolving by random chemical combinations.[13] He calculated that there are 8×10^{27} (i.e. 8 x 10 to the power of 27, or 8 followed by 27 zeros) different possible combinations of the building blocks that comprise a simple protein. In other words, there are 8×10^{27} possible combinations for these building blocks being put together, but only ONE of those combinations would create a viable protein. Even if we assume that a new combination has formed every minute that the universe has existed, after 14 billion years (the current evolutionist estimate of the age of the universe) only 4×10^{17} possible combinations would have formed.[14] The universe would have to be 10 BILLION TIMES OLDER than it supposedly is, in order to produce a single, simple protein!

Furthermore, he calculated the probability of a fully functioning cell, with its millions of interdependent parts, forming by itself to be 1 in $10^{40,000}$. (That is the number 1, followed by 40,000 zeros! There is not even a name for that impossibly large number!)

Subsequent probability estimates have been even less favourable. Biophysicist Dr Harold Morowitz more recently estimated that the chance creation of a living cell from natural processes might be 1 in 10 to the power of 10 BILLION.

Dr Thomas Heinze, Professor of Organic Chemistry at Friedrich Schiller University, Germany, in his article, "*Did God Create Life? Ask a Protein*"[15], states:

> "*Proteins are so complex they will not form anywhere in nature except in living cells. Then, if a protein is to perform its task, its production must be carefully regulated, but even then, it will not function unless it also has the correct address tag and is properly folded. All these systems would have to have been in place or the 'first cell' could not function. These systems, however, are just the tip of the iceberg. I chose them to illustrate the many coordinated systems that would have to have been present before the first cell would work. The teaching that the first cell spontaneously popped into being without the involvement of the Creator has its basis in the pre-scientific myth that single-celled creatures were simple. It obviously does not stand up under today's knowledge that a cell's*

DNA, RNA, membranes and proteins are extremely hard to make, and when proteins are made, they must be properly folded, addressed, and turned on and off at just the right times. None of these brilliant solutions could invent itself, yet <u>no 'first cell' could exist without all of them</u>. They could not have happened without a very intelligent Creator." (Emphasis mine).

The late Dr Sir Fred Hoyle stated that the single cell is so unbelievably complex that the chance of it forming from non-living matter by natural chemical processes is *"less probable than the chance that a tornado sweeping through a junkyard might assemble a fully functioning 747."*[16]

A FAMOUS CONVERSION

The growing concept of the irreducibly complex, factory-like nature of the single cell led to one of the most famous scientific conversions of the last century. Dr Dean Kenyon, Professor Emeritus of Biology at San Francisco State University, was once a convinced evolutionist who co-authored the evolutionary biology textbook, *Biochemical Predestination*.[17] In the late 1970s, however, he became increasingly aware of a variety of evidence contradicting evolution, including the inexplicable irreducible complexity of the single cell. Dr Kenyon became convinced that the theory of evolution could not explain how such an irreducibly complex system could have evolved, and he became an outspoken advocate of intelligent design and creationism. He began teaching creationism in his biology courses at university, and he became a strong proponent of creationism in various public debates and court cases.[18]

Dr Kenyon published many subsequent academic papers and books, repudiating his previously held position and advocating strong evidence for intelligent design.[19] He wrote:

"*I no longer believe that life arose spontaneously from non-living matter.*"[20]

In an interview in the 2010 Illustra Media video, "*Unlocking the Mystery of Life*", he comments on the profound impact that the irreducible complexity of the single cell had upon him, stating:

> "*This is absolutely mind boggling to perceive at this scale of size, with such finely tuned apparatus. ... We see the details of an immensely complex molecular realm of genetic information processing and it is exactly this new realm of molecular genetics where we see the most compelling evidence of design on the earth. Nothing short of an intelligence could have created this intricate cellular apparatus.*"[21]

CHARLES DARWIN'S FRANK ADMISSION

When Charles Darwin published his theory of evolution in 1859, he was acutely aware of its tenuous nature. In a statement on page 189 of his book, *On the Origin of Species*, he made this frank admission:

> "*If it could be demonstrated that any complex organ existed which could not possibly have been formed by numerous, successive, slight modifications, my theory would absolutely break down.*"[22]

Increasingly today, many scientists are claiming that we have now discovered numerous examples of the kind of irreducibly complex biological systems that Darwin stated could disprove his theory. Biochemist, Dr Michael Denton, in his book, *Evolution: A Theory in Crisis*, states:

> "We now know that there are in fact tens of thousands of irreducibly complex systems on the cellular level. Specified complexity pervades the microscopic biological world. Although the tiniest bacterial cells are incredibly small, weighing less than 10^{-12} grams, each is in effect a veritable micro-miniaturized factory containing thousands of exquisitely designed pieces of intricate molecular machinery, made up altogether of one hundred thousand million atoms, far more complicated than any machinery built by man and absolutely without parallel in the non-living world."[23]

Commenting on Darwin's frank admission regarding the possibility of an irreducibly complex system disproving his theory, Dr Michael Behe states:

> "As the number of unexplained, irreducibly complex biological systems increases, our confidence that Darwin's criterion of failure has been met skyrockets toward the maximum that science allows."[24]

One suspects that if Charles Darwin was alive today, he would concede defeat, based upon his own criterion of failure. Many of his present-day supporters, however, appear much less open to evaluating the theory in the light of contrary evidence than Darwin was, himself.

DECEPTIVE CLAIMS

"Researchers Solve Puzzle of Origin of Life on Earth". That was the headline of an article published in the online science magazine, SciTechDaily.com, in August 2020. It reports the findings of a team of researchers from the University of Washington, recently published in *The Proceedings of the National Academy of Sciences*. The problem is that the headline just isn't true. Their

research hasn't even come close to solving the puzzle of how life originated on Earth. What was their momentous discovery? Have they discovered how the first cell formed? Has the holy grail of evolution finally been uncovered?

No. All that they discovered was that by adding RNA molecules to fatty acid cell membranes, the cell membranes became more durable and less likely to deteriorate in salty ionised solutions such as might have been present on ancient Earth. To put it more simply, they took a **PRE-EXISTING** RNA molecule and added it to a **PRE-EXISTING** cell membrane and noticed that the cell membrane became more durable. They did not artificially produce ANY PART of a functioning cell, nor did they show how these components could come into existence by themselves!

At the end of the researchers' paper they stated that their next step is to try to work out how individual protein molecules could *"form and bind to each other and form functional machines"* (quoted in scitechdaily.com).

My response would be to say to them, *"Good luck with that! Because THAT is the essence of the puzzle you claim to have solved, and you have just admitted that you really haven't even BEGUN to solve it!"*

This kind of gross exaggeration and deception in public scientific announcements is, sadly, not unusual. In 2015, sciencemag.org published an article entitled, *"Researchers May Have Solved Origin of Life Conundrum"*. Without going into the complexities of that paper, it too was a complete over-exaggeration which provided NO EXPLANATION as to how a simple cell with its millions of functioning molecular machines and its unbelievably complex programming could have arisen.

. . .

SCIENTISTS CONTINUE TO BE BAFFLED

Today, at this very moment, the scientific community remains completely baffled as to how cellular life could have originated via natural means:

> "The novelty and complexity of the cell is so far beyond anything inanimate in the world today that we are left baffled by how it was achieved." (Dr M. W. Kirschner, Professor and Chair, Department of Systems Biology, Harvard Medical School, USA.).

> "We must concede that there are presently no detailed Darwinian accounts of the evolution of any biochemical or cellular system, only a variety of wishful speculations." (Dr M. Harold Franklin, Emeritus Professor of Biochemistry and Molecular Biology Colorado State University.)

> "The biggest gap in evolutionary theory remains the origin of life itself... the gap between such a collection of molecules [amino acids and RNA] and even the most primitive cell remains enormous."(Dr Chris Wills, Professor of Biology at the University of California, USA.)

> "We are as much in the dark today about the pathway from nonlife to life as Charles Darwin was when he wrote, 'It is mere rubbish thinking at present of the origin of life; one might as well think of the origin of matter.'" (Dr Paul Davies, Director of BEYOND: Center for Fundamental Concepts in Science at Arizona State University.)

Recognising the impossibility of a single cell coming into existence via chance processes, Dr Eugene V. Koonin (not a creationist) states:

> "The origin of life is the most difficult problem that faces evolutionary biology and, arguably, biology in general. Indeed, the problem is so hard, and the current state of the art seems so frustrating, that some researchers prefer to dismiss the entire issue as being outside the scientific domain altogether, on the grounds that unique events are not conducive to scientific study."[25]

In other words, Dr Koonin is saying that *'we have no idea how life could have originated; the problem is too difficult to solve, so some of us are just choosing to ignore it!'* He also appears to be admitting that the answer to the origin of life must lie BEYOND the realm of science and naturalistic causes.

The impossibility of abiogenesis – of life arising via chance processes, has led some evolutionists to propose some ridiculous theories. For instance, Dr Richard Dawkins, in an interview with American journalist and commentator, Ben Stein, was pushed to explain how life could have originally started on earth. After conceding that there was no known natural biological process whereby life could have possibly formed, he made the following extraordinary statement:

> "It could be that, at some earlier time, somewhere in the universe, a civilisation evolved, by probably some kind of Darwinian means, to a very, very high level of technology, and designed a form of life that they seeded onto, perhaps, this planet."[26]

The idea that aliens created life on earth, however, raises two very obvious questions: Where are they? And, more importantly, who created *them*? Suggesting aliens as the originators of life on earth merely pushes back the question of ultimate origins by an additional step. It does not solve the problem of how *any* life can evolve from non-living matter *anywhere* in the universe!

THE ORIGIN OF LIFE EXPLAINED IN THE BIBLE

Of course, there is a very simple and obvious explanation for the origin of life. This explanation has actually been around for a very long time, and it makes perfect sense. You can find this explanation in the first verse of the Bible:

> "In the beginning, God created the heavens and the Earth."
> (Genesis 1:1)

There have always been scientists who have agreed with the Bible's portrayal of God as the author of life. Louis Pasteur (1822 – 1895) was a biologist of the highest order who specialised in studying the microbiological world. Known as the co-founder of modern microbiology and immunology, and the developer of vaccination and pasteurisation, he once stated,

> "The more I study nature, the more I stand amazed at the work of the Creator. Science brings men nearer to God."[27]

The evidence of the irreducible complexity of the living cell and the widely-recognised impossibility of life arising through random chemical processes cry out for an intelligent, all-powerful Creator. The mystery of the origin of biological life is stunning evidence for the existence of God.

5

EVIDENCE 3: INTELLIGENT DESIGN

The concept of intelligent design has received much attention in recent years. As scientists have grown in their understanding of the complex forces and structures undergirding our universe, it has become increasingly less likely that the universe is the product of brute chance. Even among those scientists who refuse to believe in a personal God (a God with a mind and personality), many are having to concede that some form of intelligence was at work in the design and formation of these fundamental forces and structures.

Walter Bradley, a retired professor of mechanical engineering at Texas A&M University, a towering figure within the intelligent design movement, and co-author of *The Mystery of Life's Origin*, stated:

> "It is quite easy to understand why so many scientists have changed their minds in the past thirty years, agreeing that the universe cannot reasonably be explained as a cosmic accident.

> *Evidence for an intelligent designer becomes more compelling the more we understand about our carefully crafted habitat.*"[1]

Evidence of intelligent design is everywhere, but in this chapter, we will limit our discussion to four examples:

- Intelligent design evident in the fine tuning of universal constants

- Intelligent design evident in planetary habitability factors

- Intelligent design evident in DNA

- Intelligent design evident in the mystery of human consciousness

INTELLIGENT DESIGN EVIDENT IN THE FINE TUNING OF UNIVERSAL CONSTANTS

The fundamental forces that underpin our universe are many and complex. These are the forces that hold the very fabric of our universe together, without which, biological life would not be possible and, in the case of some constants, matter itself could not exist. These forces are called universal or physical "constants", because they do not change, and they are incredibly fine-tuned to be precisely what is necessary for a life-permitting universe to exist. There are 34 recognised universal constants, from the tiny elementary charge that holds atoms together to the cosmological constant that keeps the universe from flying apart or from collapsing in upon itself.

Table of Universal Constants

Quantity	Symbol	Numerical value	Unit
Acceleration of free fall (standard)	g_n	9.8066	m/s^2
Atmospheric pressure (standard)	p_0	1.0132×10^5	Pa
Atomic mass unit	u	1.6606×10^{-27}	kg
Avogadro constant	N_A	6.0220×10^{23}	mol^{-1}
Bohr magneton	μ_B	9.2741×10^{-24}	J/T, A m^2
Boltzmann constant	k	1.3807×10^{-23}	J/K
Electron			
charge	$-e$	1.6022×10^{-19}	C
mass	m_e	9.1095×10^{-31}	kg
charge/mass ratio	e/m_e	1.7588×10^{11}	C/kg
Faraday constant	F	9.6485×10^4	C/mol
Free space			
electric constant	ε_0	8.8542×10^{-12}	F/m
intrinsic impedance	Z_0	376.7	Ω
magnetic constant	μ_0	$4\pi \times 10^{-7}$	H/m
speed of electromagnetic waves	c	2.9979×10^8	m/s
Gravitational constant	G	6.6732×10^{-11}	N m^2/kg^2
Ideal molar gas constant	R	8.3144	J/(mol K)
Molar volume at s.t.p.	V_m	2.2414×10^{-2}	m^3/mol
Neutron rest mass	m_n	1.6748×10^{-27}	kg
Planck constant	h	6.6262×10^{-34}	J s
normalised	$h/2\pi$	1.0546×10^{-34}	J s
Proton			
charge	$+e$	1.6022×10^{-19}	C
rest mass	m_p	1.6726×10^{-27}	kg
charge/mass ratio	e/m_p	0.9579×10^8	C/kg
Radiation constants	c_1	3.7418×10^{-16}	W m^2
	c_2	1.4388×10^{-2}	m K
Rydberg constant	R_H	1.0968×10^7	m^{-1}
Stefan-Boltzmann constant	σ	5.6703×10^{-8}	J/(m^2 K^4)
Wien constant	k_w	2.8978×10^{-3}	m K

These recognised universal constants are so finely tuned that altering one by even the tiniest fraction of one percent would render life and, in the case of some of the constants, the existence of matter itself, impossible.

Take, for example, the elementary charge, also known as the electron charge or the strong nucleic force. This is the attractive force that exists between electrons and protons within an atom, which has been determined to be $1.602176634 \times 10^{-19}$ C (coulombs). This tiny force is precisely what is required to keep electrons in stable orbit around the nucleus of an atom. If the elementary charge was even one billionth of one percent greater, electrons would collapse into the nucleus of the atom. If the charge was only one billionth of one percent weaker, electrons would no longer be held in stable orbit and would fly away from the nucleus. In either case, physical matter would simply cease to exist.

Gravity is another example. This is the attractive force that matter exerts upon matter. It is the force that holds us to the surface of the earth. It is the force that holds our planet together. Gravity is the force that allows planets and stars to form and remain a stable mass. It is a fundamental force that holds the fabric of our universe together. The force of gravity has been measured to be 6.6732×10^{-11} Nm^2 kg^{-2}. This is a surprisingly small force. Let me expand it out so you can see how tiny it is: 0.000000000066732 Nm^2 kg^{-2}. This is why, when you sit next to someone, you cannot feel the gravitational attraction between you and the other person (although, depending on the other person, you may feel a different kind of attraction!). Gravity only becomes noticeable when at least one of the objects in the equation is very massive - like a planet. The gravitational constant is extremely finely tuned. Dr Robin Collins, Professor Emeritus of Theoretical Physics at North Western University, states:

> "If the gravitational constant was increased by just one part in ten thousand billion billion billion, that small adjustment would increase gravity by a billion-fold!"[2]

In such a case all matter would collapse in upon itself and the universe would consist of nothing but black holes. Conversely, if the gravitational constant was decreased by a similarly minute amount, matter would simply drift apart. There would be no stars, no planets, no basis for physical life.

The so-called weak nucleic force is a physical constant which operates within the nucleus of an atom. It is so finely tuned that altering its value by even one part in 10 to the power of 100 would prevent the very existence of matter.

Similarly, the cosmological constant, the energy density of "empty" space, is even more precisely fine-tuned than gravity. It needs to be precisely at its current value in order for the universe to exist at all. Dr William Lane Craig explains:

> "A change in the value of the so-called cosmological constant, which drives the acceleration of the universe's expansion, by as little as

one part in 10 to the power of 120 would have rendered the universe life-prohibiting."³

To give you an idea of how staggeringly small these numbers are, and therefore, how extraordinarily narrow is the degree of this fine tuning, consider the following numbers. If the universe is 14 billion years old (as evolutionists suggest), this equates to 4.4 x 10 to the power of 17 seconds (44 followed by 17 zeros). The number of atoms in the entire universe has been estimated to be 10 to the power of 80.⁴ That's 10 with 80 zeros after it. These are incomprehensibly huge numbers. And yet the fine tuning of many of the universal constants involves precision down to ONE PART in quadrillions of times larger numbers than these!

This incredible fine tuning applies to ALL the physical constants. Each of them is precisely tuned to an infinitesimally exact strength that is essential for the universe to exist and for life to be possible. To change any of them by even the most miniscule of fractions would render the universe as we know it no longer possible.

The possibility of these universal constants all arriving at their precise values by sheer chance alone is so infinitesimally small as to be an impossibility. Given that each of these forces could have been formed at *any* possible value, from extremely strong to extremely weak, the chance of them *all* arriving at precisely the exact strength necessary for a life-permitting universe is simply astonishing! Professor of Philosophy, Dr Robin Collins states:

> "The chance of just two of these cosmological constants developing by sheer chance, is one in 100 million trillion trillion trillion trillion trillion trillion. That's more than the number of atoms in the universe! And that's just TWO of the constants!" ⁵

Dr Collins' probability estimate, above, relates to just two of the constants. The probability of all 34 constants spontaneously appearing at their precise strengths simultaneously is calculated to be about 1 in 10 to the power of 600 (10 with 600 zeros after it)! To give you an idea of how impossible that number is, it is the number of atoms in our universe multiplied by 10 to the power of 520 further universes! In other words, it is a complete statistical impossibility that all these precisely tuned constants could have developed by chance.

Dr Paul Davies, Professor of Theoretical Physics, Adelaide University, states,

> "The physical universe is put together with an ingenuity that is so astonishing, with physical constants that are so impossibly perfect, that I can no longer accept it as the product of brute chance".[6]

Dr Fred Hoyle, astrophysicist and mathematician, Cambridge University, stated,

> "A common-sense interpretation of the facts suggests that a super-intellect has monkeyed with physics, as well as with chemistry and biology, and that there are no blind forces worth speaking about in nature. The numbers one calculates from the facts seem to me so overwhelming as to put this conclusion almost beyond question."

Albert Einstein, in a letter to a student, Phyllis Wright, on January 24, 1936, wrote:

> "Everyone who is seriously involved in the pursuit of science becomes convinced that some spirit is manifest in the laws of the universe, one that is vastly superior to that of man. In this way the pursuit of science leads to a religious feeling of a special sort, which

is surely quite different from the religiosity of someone more naïve."[7]

INTELLIGENT DESIGN EVIDENT IN PLANETARY HABITABILITY FACTORS

Similar to the fine-tuning of physical constants, the position of the earth within the solar system and within the galaxy has been specifically designed to afford the perfect conditions for biological life. The earth is the perfect distance from the sun. It orbits the sun at the optimal, life-permitting distance, in what is referred to as the circumstellar habitable zone, or the "Goldilocks zone", meaning that it is "just right" for biological life. A little closer, and all water would evaporate; a little further away and we would be a frozen planet.

The earth's size, mass, rate of spin and axis of tilt (23.5 degrees) are also absolutely perfect for biological life to survive and flourish. Our atmosphere has the perfect combination of gases necessary for life. The earth's magnetic field is precisely the right strength, acting as a shield, protecting us from the majority of harmful radiation from the sun.

The moon also plays a vital role in creating the tidal forces necessary for circulating and oxygenating the earth's water systems. At around one-eightieth the mass of the earth, the moon is exceptionally large for the size of the earth when compared to the 60 moons of other planets in our solar system. This ratio of moon to planet is unique among all the planets with moons that have been discovered in our galaxy, yet this is the perfect size for the life-giving function that it fulfils for our planet. The moon's strong gravitational force upon the Earth's oceans and tectonic plates is essential for circulating nutrients, generating ocean currents and maintaining an even surface temperature, as well as perpetuating our seasons by moderating our planet's axial 'wobble'.

The sun, too, is just the right kind of star; a yellow dwarf, main sequence star. It is the right size, the right temperature, the right spectral class and the right mass. Only about 7% of stars in the observable universe would be right for us.

The location of our solar system within the Milky Way galaxy is also remarkably ideal. If we were closer to the galactic core, the huge amount of radiation would render life as we know it impossible. Our location on the outer edge of a spiral arm of the galaxy keeps us at a safe distance from the maelstrom of deadly radiation at the galactic core. Propitiously, it also places us in an ideal position to observe the universe. If our solar system was embedded more deeply in the densely populated galactic core, all we would see when we looked up into the night sky would be a thick, impenetrable carpet of stars. Our ideal location on the outer edge of the galaxy allows us to not only observe the shape and nature of our own galaxy from a wonderful vantage point (side on), but also enables us to look away from our own galaxy, in the opposite direction, and observe billions of other galaxies in the rest of the universe. It is as if our planet was deliberately placed in the perfect position to not only keep us safe, but also to enable us to discover the universe for ourselves.

Another fascinating "coincidence" is our ability to observe solar eclipses. These occur when the moon passes directly in front of the sun, momentarily eclipsing it. These solar eclipses have enabled scientists to study the sun's corona (outer atmosphere) in a way that would otherwise be impossible. Solar eclipses have also led to some startling scientific discoveries regarding the nature of the sun and the properties of light, as scientists have been able to observe the refraction of light around the sun. In fact, it is through observations of solar eclipses, that scientists have made many important discoveries about the nature of our universe. For example, during an eclipse, we are able to glimpse stars that are behind the sun, as the sun's gravity bends the light from those stars around itself, making them visible to us. By observing the ability of gravity to bend light in this way, scientists were able to confirm Einstein's General Theory of Relativity.

What makes solar eclipses so special, is how precisely the position and size of the earth, moon and sun need to be in order for an such eclipse to be possible. What is extraordinary is that, when viewed from the perspective of earth, the moon appears to be *exactly* the same size as the sun, so that when it passes in front of the sun it *exactly* covers the sun, while leaving the sun's outer atmosphere visible. For about two minutes during an eclipse, the moon covers the intensely bright photosphere of the sun, enabling us to observe its thin faint chromosphere and the spectacular corona with its dramatic prominences. This is because the sun is 400 times bigger than the moon, but it is also 400 times further away. In fact, the ratio of the size of the moon when compared to the size of the sun is **exactly the same**

ratio as their distances from earth! This is an extraordinary "coincidence" which baffles atheistic scientists. The likelihood of this astronomical arrangement arising by sheer chance is too astonishingly small to be considered possible. It is as if "Someone" deliberately designed our earth and our solar system to enable us to conduct scientific investigations into the nature of the universe.

Of course, evolutionists love to talk up the probability of the existence of earth-like, habitable exoplanets (planets outside our solar system). They would love to prove that life is not as miraculous as creationists propose. In recent years there have been frequent "announcements" from astronomers estimating that there may be anywhere from 10 million to 100 billion habitable earth-like exoplanets in the universe. Significantly, of the several thousand exoplanets discovered to date, none of them are even remotely earth-like or habitable. The Planetary Habitability Laboratory website is run by the University of Puerto Rico in conjunction with data from the Kepler and Hubble space telescopes. It keeps an up-to-date list of potentially habitable planets. Currently they list ten exoplanets that "may" be the right distance from their sun. The only one remotely similar in size to earth, Proxima Centauri B, is twice the size of earth, is tidally locked to its sun (one face permanently facing the sun) and orbiting a red dwarf star (the wrong kind of star). Wikipedia's entry for Proxima Centauri B, states:

> *"Its habitability has not been established, though it is unlikely to be habitable since the planet is subject to stellar wind pressures of more than 2,000 times those experienced by Earth from the solar wind."*[8]

What an extraordinary statement! *"Its habitability has not been established*"? Really? I would say its **lack** of habitability has been firmly established.

Recently, I watched the first episode of a new scientific documentary series on cosmology called "The Great Acceleration" (produced by Wildbear Entertainment) and was shocked at the bald-faced lies that were presented to prop up an atheistic view of the universe. At one point, the narrator, Dr Shalin Naik, said, *"Planets like our own are everywhere."* Shortly after, Dr Brad Tucker, an astrophysicist at the Australian National University, boldly proclaimed, *"We are not unique. Kepler* [a space telescope launched in 2009] *told us that there are 20 billion planets like Earth in our galaxy alone."*

According to this documentary, there are 20 billion Earth-like planets in our galaxy!

Before I explain why atheistic scientists are so keen to paint Earth as unremarkable and common, let me firstly reveal how shocking and blatant this exaggeration is, by pointing you to the facts. Because the *fact* is that scientists have not yet discovered a SINGLE planet that is even *remotely* similar to Earth.

As of 24 August 2020, scientists have discovered a total of 4,201 confirmed exoplanets (planets orbiting other stars) and NONE of those is a confirmed Earth-like planet capable of sustaining life. This is a LONG way from 20 billion! You can easily verify this information on a plethora of scientific websites. For example, NASA has a website called "Exoplanet Catalogue" and there is another website called "Open Exoplanet Catalogue". These and many other websites list each confirmed exoplanet and provide data concerning the planet's mass, radius, distance from its star and any other characteristics that have been discovered such as possible atmospheric gases, temperature range, etc.

Because of the vast distances from Earth of these exoplanets, very few of them can actually be viewed by even our most powerful space telescopes: Hubble and Kepler. Almost all of these planets have simply been *inferred* by observing tiny eccentric perturbations in the movement of their relevant stars, supposedly caused by the gravitational effect of a probable planet in orbit. Scientists observe the faintest wobble in a star's trajectory and deduce the existence of a planet, estimating its probable mass and distance from the star. About 97% of all exoplanets discovered so far have been detected via this method and are too distant to be viewed at all.

A handful of exoplanets, however, are close enough to be viewed with the Hubble or Kepler telescopes. But even in these cases, using the most powerful magnification available to us, they are merely tiny points of light, reflecting their star's brilliance. The planet itself cannot be seen in any detail. In the case of these few planets, spectral analysis of the light being reflected by the planet enables scientists to estimate the kind of gases that might be in their atmosphere, if they have an atmosphere at all. So far, no planets with a breathable, Earth-like atmosphere have been discovered.

Furthermore, the vast majority of exoplanets orbit their star OUTSIDE the habitable zone of their star. They either orbit too near (and are thus too hot) or too far from the star (and are too cold) to be life-permitting. Most exoplanets are also far too massive to sustain life as we know it. Of the 4,021 currently known exoplanets, only 20 are Earth-sized rocky planets that orbit within their star's habitable zone. Even in these cases, scientists have no idea whether these planets have an atmosphere, liquid water or any of the other essential characteristics necessary for life.

In fact, cosmologists have a long list of characteristics deemed to be essential if a planet is to be life-permitting:

- Orbiting within the star's habitable zone

- Liquid water

- Breathable atmosphere

- Moving tectonic plates

- Liquid core

- Rocky planet (not a gas planet)

- Magnetic shield (to protect from harmful radiation)

- Rotates (is not tidally locked with one face permanently facing its sun)

- Axial tilt for seasons

- Elliptical orbit for seasons

- Large moon for tidal circulation of nutrients and temperature control

- Temperature range between 0 and 100 Celsius (so liquid water can exist)

- Episodic volcanism and out-gassing

- Life-permitting atmospheric surface pressure

- Life-permitting rotation rate

- Orbiting the right kind of star (as many star types would be deadly to life)

- The presence of carbon-based photochemistry

- The right distance from the galactic core (too close would be deadly because of the intense radiation in that densely packed region of space)

I'll stop there, because scientists currently propose about 30 essential factors that are necessary for a planet to be habitable and life-permitting. Each of these criteria is essential; without any one of these, the development and continued viability of biological life would not be possible. So far, of the 4,021 exoplanets currently discovered, NONE of them comes even close to meeting all these criteria. For example, we have not yet discovered a single planet with liquid surface water. And we have not yet discovered a single planet with a breathable atmosphere. Not one!

So, how do scientists like Dr Brad Tucker get away with claiming that *"there are 20 billion planets like Earth in our galaxy alone"*? There are two parts to the answer.

Firstly, the "20 billion" figure, is an estimate of the existence of ANY kind of planet, based on current findings. Currently, about 1 in 10 stars in our nearby galactic neighbourhood appear to have at least one planet. Given that our Milky Way Galaxy has about 200 billion stars, scientists therefore extrapolate that our galaxy possibly contains about 20 billion planets. So far, so good. But the 'like Earth" claim, is a complete fabrication. As we have already seen, NONE of the exoplanets so far discovered is even remotely like Earth, as far as we have been able to tell. Significantly, it is only in popular media and documentaries produced for mass consumption where this kind of blatant exaggeration is foisted upon the general public; this kind of outrageous claim is never published in formal scientific journals.

Probability estimates have been made to ascertain the likelihood of all 30 habitability factors coming together in the same planet by pure chance, and it turns out that there aren't enough estimated planets in the entire universe for an Earth-like planet to have developed by chance. A recent study by astrophysicist, Dr Erik Zackrisson from Uppsala University, Sweden, used computer modelling to simulate the universe's formation following a Big Bang. His modelling estimated that the probability of an Earth-like planet forming by pure chance is 1 in 700 quintillion (1 in 7 with 20 zeros after it).[9] This is essentially zero in statistical terms. His landmark study was published in *The Astrophysical Journal* and poured cold water on the exaggerated claims of atheistic, evolution-believing scientists.

So, why are atheists and evolutionists so keen to claim that Earth is not unique? The reason is that IF they can show that planets like Earth can form purely by chance and in great numbers via mechanistic evolutionary processes, then the necessity for a divine creator is done away with. Effectively, atheistic scientists are saying to the general public, *"Nothing to see here, folks! There's nothing special about the Earth. Mother nature is throwing up planets like this all over the place. There is no God who has created a special home for us."*

Increasingly, however, their deliberately misleading claims, are being refuted by the evidence of hard science. The Earth is unique. It is a shining jewel in the vast bleakness of the universe. It is a miraculous place; a safe haven in the harshness of our beautiful but deadly universe. The Earth stands alone as a work of wonder with a statistically impossible confluence of habitability factors that absolutely screams out that there is an intelligent, all-powerful Creator.

It is disappointing that atheistic scientists repeatedly get away with making their exaggerated claims via popular media. And, sadly, the vast majority of the general public simply lap it up as one more piece of evidence confirming the concept that we are all just the product of brute chance. But the truth is that the more we discover about the universe, the more we realise how special and miraculously unique our planet really is.

In 2015, John Horgan, a deist rather than a Christian, wrote an article in "*American Scientific*":

> "*The more scientists investigate our universe, the more improbable our existence seems. If you define a miracle as an infinitely improbable event, then our existence, you might say, is a miracle. Scientists try in vain to hand-wave our improbability away ... My own mystical intuitions keep me from ruling out the possibility of supernatural creation.*"[10]

Sir Isaac Newton (1643 – 1727), who formulated the laws of motion and universal gravitation, was convinced that the universe's obvious intelligent design and the Earth's unique place within it was proof of God's existence. He stated:

> "*This most beautiful system of the sun, planets, and comets could only proceed from the counsel and dominion of an intelligent Being. This Being governs all things, not as the soul of the world, but as Lord over all; and on account of his dominion he is wont to be called "Lord God" or "Universal Ruler". The Supreme God is a Being eternal, infinite, and absolutely perfect.*"[11]

INTELLIGENT DESIGN EVIDENT IN DNA

In examining the evidence for intelligent design within the physical universe, we now put aside the telescope and take hold of the microscope. For deep inside the nucleus of almost every living cell is an extraordinarily complex set of operating instructions that can only have been placed there by a supremely intelligent mind.

DNA (deoxyribonucleic acid) is the "computer coding" that resides within the nucleus of almost every cell in your body. (I say 'almost' every cell, because there are a few types of cells – mature red blood cells, and cornified cells in skin, hair and nails – which start their life with DNA but their DNA is deleted when the cell reaches maturity). DNA is the complete instruction manual of how to build every component of your body; how to repair it when bits break and how to ensure its ongoing proper functioning. It is a vastly complex piece of code. It contains 3.2 billion "letters" which are combined to spell out very specific genetic instructions, built into a biological construct in the shape of a spiralling double helix.

To give you an idea of how much information is in human DNA, the Encyclopedia Britannica has 200 million letters across 32 volumes. Human DNA, by comparison, has 3.2 *billion* pieces of information (pairs of 'letters'). This is the equivalent of approximately 16 sets of the 32 volume Encyclopedia Britannica! To put it another way, if DNA could be printed as words in a book, it would fill up over 20,000 copies of the book you are currently holding in your hand (or looking at on your screen)!

The DNA in the nucleus of each cell is an extremely fine molecular structure, folded in upon itself, over and over again. If you could unfold your DNA into a straight line, the DNA in a *single cell* would be about 2 metres long (6 feet)[12] - although you would need an electron microscope to see it, because it would only be a single molecule in width. That is the length of the DNA in just *one* cell. Your body has about 10 *trillion* cells. If you could unfold all the DNA from every cell and place it end to end, your DNA would stretch out to 1.2 million kilometres (744 million miles)! That would stretch from the earth to the sun and back again, *4 times*!

What is most relevant to our discussion of intelligent design, however, is the incredibly complex nature of DNA and the vast amount of information it contains.

The Human Genome Project was an international research project aimed at mapping the exact sequence of base pairs (or "letters") within the entire human DNA. It was an enormous task and remains, to this day, the largest collaborative biological project ever undertaken. Starting in 1990, using the most sophisticated computers available and involving hundreds of geneticists from around the world, it took until 2003 before the entire DNA sequence was mapped.[13] It took *13 years*, with the best available computers running 24/7, to determine the exact order of all 3.2 billion base pairs within our DNA. That is how vast and complex the coding is within our DNA!

Although we now know the exact order of all the "letters" (base pairs) within our DNA, geneticists only understand the nature and function of *less than 1%* of those biological instructions.[14] We have mapped our DNA, but we have very little understanding of what most of it "says". It is a highly complex code. Even today, with our most sophisticated computers and technology, we have barely begun to understand the complex code of our DNA.

Now we come to the key question: Where did this vast amount of complex information come from? Who wrote the code and placed it inside every cell of our body? The evolutionary story has no explanation for this. No amount of random mixing of chemicals or stirring of the primordial swamp could conjure up a process whereby this vast amount of complex coding simply popped into existence.

In case you are thinking that the earliest, simple lifeforms would have needed a much smaller amount of DNA, consider the following. According to the U.S. National Science Foundation,[15] our human DNA contains about 23,000 genes (a gene is

Evidence 3: Intelligent Design | 81

a section of DNA, containing a whole "page" or "chapter" of instructions regarding a particular function or organ). By comparison, a water flea (*Daphnia pulex*) has 31,000 genes![16] The humble earth worm has 20,000 genes![17] Even more astonishing is the complex DNA chain found within single-cell organisms. Compared to human DNA, which has 3.2 billion base pairs, the single-cell *amoeba proteus* has 290 billion base pairs and the single-cell *amoeba dubia* has 670 billion base pairs![18] In other words, amoeba DNA is *90 times* longer than human DNA and protozoa DNA is *209 times* longer! Scientists have no understanding of why this is so,[19] but one thing it demonstrates very clearly is that there is no such thing as a "simple" organism.

Dr Macki Giertyche, Professor Emeritus of Genetics, at Torun University, Poland, states:

> *"The science of genetics makes it clear that at no time in the past can there have been such a thing as a simple organism. All organisms, however primitive they may appear, are complex and bursting with information. And we know that this information must have been there from the very beginning. For example, the very complex DNA & RNA protein replicating system in the cell must have been perfect from the very beginning - if not, life could not exist. The only logical explanation is that this vast quantity of information came from intelligence. Every bacterium, every microscopic cell, is so precisely programmed that we have to assume that the information contained in them comes from an intelligence far beyond our own.... Evolutionists have no idea how this information system is produced."*[20]

Evolution teaches that life began with very simple, single-cell organisms and gradually evolved into more complex life-forms. But our current understanding of genetics reveals that even the smallest of organisms is filled with a vast amount of intelligent, highly detailed coding that could not possibly have arisen by chance. No amount of lightning strikes into a primordial swamp could instantly (or ever) create a series of biological instructions equating to, in the case of *amoeba dubia*, **15,000 sets** of the 32 volume Encyclopedia Britannica!

Dr John Ashton states, in his book, *Evolution Impossible*:

> "To date, I have found no reputable published scientific paper that explains a proven mechanism for how this huge amount of highly complex genetic information could arise by chance. Nor can I find any scientific papers reporting the observation of new meaningful genetic information arising by chance."[21]

What is also remarkable about DNA is the ability of individual cells to decode and read the portion of the DNA that is applicable to their own function and ignore the rest of the information in the DNA. In the case of humans, almost every cell in our body has a complete strand of DNA stored in its nucleus- a complete instruction manual for the building, repairing and proper functioning of the whole human body. Yet, remarkably, each cell only accesses that part of the DNA that is applicable to itself and disregards the rest. Although the world's most gifted scientists have only been able to decode less than 1% of our DNA (they have almost no idea what the rest of it means - what each segment is for), our cells can read it easily, and can extract the necessary instructions and follow them.

Who programmed our cells to be able to do this? How does an eye cell know which part of the DNA to access and read? How does a liver cell know where to look in the instruction manual to repair itself? This extraordinary ability of our cells is another aspect of our biological sophistication and complexity that cannot be explained by random evolution and which cries out for an Intelligent Designer.

DNA and Irreducible Complexity

In order for a cell to be alive and able to reproduce, it requires its complete set of DNA. This genetic coding cannot have accumulated over vast amounts of time, because without a full set of DNA, with a complete set of genes to regulate all of the cell's functions, the cell would not be alive and would not be able to replicate and pass on its genetic information to subsequent generations.

The minimum requirement for a single cell to be a living, functioning, reproducing entity, is a *full set* of genetic coding, with all the relevant genes, to be present *from the very beginning*. In this sense, the DNA within a cell is irreducibly complex. For example, the previously mentioned single-celled *amoeba dubia* needs the complete DNA of 670 billion base pairs to be present in order for it to exist. An *amoeba dubium* with only 300 billion base pairs would be incomplete and unviable. It would not be a living amoeba at all, but merely a collection of chemicals and molecules. The full strand of DNA had to come into existence *in its completed form* in order for any cell to be alive and able to reproduce. Gradual, incremental accumulation of DNA information is impossible without a living organism to reproduce and pass on that incremental information. And a living organism cannot exist without a full set of DNA. In the words of Joseph Heller's famous book, it's a "Catch 22" situation![22]

A recent article on the website, Creation Science Hall of Fame, states that if the entire DNA sequence of an *amoeba dubia* was printed in standard font size, it would result in enough books to fill the U.S. National Library of Congress 10 times over![23] How could that vast amount of genetic information have come into existence simultaneously and by chance?

Charles Darwin, of course, knew nothing of DNA when he published his evolutionary theory in 1859. DNA was not properly understood until 1953.[24] Had Darwin and his contemporaries understood the complex nature of DNA, they would most likely never have entertained the concept of life popping into existence by chance natural causes.

DNA and Specified Complexity

As well as irreducible complexity, DNA also demonstrates specified complexity which cannot be adequately explained by evolution. Specified complexity refers to the existence of a highly complex structure or set of information that is required to exist in a very specifically defined sequence in order to be viable. The longer and more complex the sequence of information, the less likely it is that it came about by chance processes.

Some atheist scientists try to argue that the vast amount of information in the DNA strand could have developed by chance. They argue that a monkey tapping away at random on a computer keyboard would eventually correctly type not only words but also phrases, sentences, and even whole books.[25] So, given enough time, they argue that chance alone could produce the necessary genetic information for anything - whales, monkeys, humans.

Dr Julian Huxley used this 'monkeys typing randomly' argument in his debate with Rev Samuel Wilberforce in 1860 to explain how biological complexity could arise by chance given enough time. He claimed that given sufficient time, "apes" (as he called them) typing randomly could eventually type out the complete works of Shakespeare.[26]

More recently, Dr Richard Dawkins used a modern iteration of the 'monkeys typing randomly' argument in his book, *The Blind Watchmaker*. He claimed that monkeys typing randomly for millions of years would eventually type one of Shakespeare's sonnets, and he used this analogy to explain how DNA could have arisen by chance over a long period of time.[27]

At first glance, this argument may appear to have merit: Chance + Sufficient Time = Orderly Information.

Let us examine this argument logically. As there are 101 keys on a typical computer keyboard, the chance of a monkey randomly typing the six-letter word *"peanut"* is 1/101 x 1/101 x 1/101 x 1/101 x 1/101 x 1/101, which equals 1 chance in 1.06 trillion. The mathematics of probability tells us that typing at 3 letters per second, the monkey would have to type for 10,379 years to be guaranteed to produce the word "peanut"! What about the phrase *"peanuts and computers"*? To correctly type these 21 letters and spaces by chance alone, it would take 1 billion monkeys typing randomly for one thousand million million million million million years! That is trillions of times longer than evolutionists believe the universe has existed! But there aren't just 21 letters in our DNA; there are 3.2 **billion** pairs of letters ("base pairs")!

These simple probability calculations reveal how utterly ridiculous is the claim that the vast amounts of information in DNA could have come into existence by random natural processes. It is also worth noting that the above probability calculations are based on the concept of information accumulating gradually, via incremental steps over time. But we know that this could not have occurred in the case of DNA. For a cell to be a viable, living, functioning, reproducing entity, the **entire** DNA chain had to come into existence **simultaneously** and in **precisely** the correct sequence. No amount of proverbial monkeys or random processes can explain how this could have occurred.

The incredibly complex coding at the heart of every living cell, with 3.2 billion pieces of specific information, MUST have been created by an intelligent mind. No serious scientist would argue that random molecular processes could write the latest operating software for an iMac computer, yet human DNA is millions of times more sophisticated than that! Only an extremely intelligent mind could have designed DNA.

Biologist Dr Stephen Meyer, in an interview with journalist and author Lee Strobel, stated:

> "Information is the hallmark of a mind. And purely from the evidence of genetics and biology, we can infer the existence of a mind that's far greater than our own — a conscious, purposeful, rational, intelligent designer who's amazingly creative."[28]

INTELLIGENT DESIGN EVIDENT IN THE MYSTERY OF HUMAN CONSCIOUSNESS AND COGNITION

A final indicator of intelligent design which is proving to be extremely problematic for scientists who are seeking to explain life apart from any sort of transcendent Creator, is the mystery of human consciousness and cognition. These two facets of the human mind continue to defy explanation in any deterministic way via any laws of chemistry or physics. Allow me to deal with each of these two aspects separately.

Consciousness

How can a bunch of living cells, when clumped together to form a brain, give rise to consciousness – to our ability to think and reason and be self-aware? How can the chemical reactions and electrical impulses inside the brain result in consciousness? How can molecules be capable of thought? Once again, atheists and evolutionists have no viable explanation for this. The origin and source of consciousness is arguably even more problematic than the origin of the single cell.

Evolutionist philosopher, Dr Michael Ruse, FRSC, puzzles over this very problem:

> "Why should a bunch of atoms have thinking ability? Why should I, even as I write now, be able to reflect on what I am doing? And why should you, even as you read now, be able to ponder my points? ... No one, certainly not the Darwinianist [evolutionist], seems to have an answer to this. The point is that there is no scientific answer."[29]

Ruse is entirely correct. Science simply *cannot* explain human consciousness. Dr Michael Polanyi, arguably the foremost philosopher of science in the 20th century, wrote:

> "Mental processes cannot be explained by physics and chemistry. The laws of physics and chemistry do not ascribe consciousness to any process controlled by them: the presence of consciousness proves, therefore, that other principles than those of inanimate matter participate in the conscious operations of living things."[30]

Evolutionist scientists, in an attempt to explain how consciousness can arise from the human brain, have recently begun using several nebulous phrases:

Non-local consciousness: The idea that our brains are merely the conduit for the mind, not the source of its origin.[31] Scientists who are exploring this concept are investigating what they refer to as *"fringe phenomena"*, such as near-death experiences and precognition, in the hope of gaining an insight into consciousness.

Integrated Information Theory: The idea that every molecule of biological life has a tiny amount of consciousness which, when combined together produces a higher level of combined consciousness. Some proponents of this theory even suggest that *all* matter, even *non-living* matter, has a tiny spark of consciousness.

Emergent Property theory: This follows on from the previous idea. It proposes that the latent consciousness within all matter is activated by the particular complex arrangements of the human brain, which is able to activate and focus this consciousness more effectively than other animals or organisms.

What is striking about all of these theories is their frank admission that consciousness cannot be explained by mere physical chemicals and atoms. Each of these theories rests on the premise that consciousness has a non-physical source. Where these theories all fall down, however, is their inability to explain how a purely physical, mechanistic universe could have created a universal consciousness that somehow pervades each molecule of matter. Where did this consciousness come from? Evolutionists have absolutely no reasonable answer.

Human consciousness indicates that we are more than mere physical machines. We are more than flesh and blood. There is an essence within each person that transcends the merely physical. Our consciousness, our ability to think, and feel, and reason, is extremely convincing evidence for the existence of an Intelligent Designer who has implanted a mind within each of us.

Theologian and philosopher, Dr J. P. Moreland, states:

> *"You can't get something from nothing. If the universe began with dead matter having no consciousness, how, then, do you get something totally different - conscious, living, thinking, feeling, believing creatures - from materials that don't have that? But if*

everything started with the mind of God, we don't have a problem with explaining the origin of our mind."[32]

Cognition

A second, but closely linked, aspect of the human mind is cognition. By this I mean our ability to decode the fundamental laws of the universe. This involves the embracing of abstract concepts. Any animal can observe an apple falling from a tree, but only the human mind can perceive the abstract concept of gravity and seek to define and explain it in precise mathematical terms. Roy Williams, in his book *God Actually*, makes the point that there is no good explanation of this ability to engage in abstract thinking in terms of mere 'survival value'.[33] As he states,

> *"It ought to be enough that Man, like many other animals, is capable of seeing and dodging out of the way of the falling apple. Anything else looks like a huge case of overkill."*[34]

Our ability to discover and embrace theoretical knowledge that has no obvious survival value argues against its origin via natural selection.

Dr Thomas Nagel, Professor of Philosophy and Law, Emeritus, at New York University, made just this point:

> *"Darwinism may explain why creatures with vision or reason survive, but it does not explain how vision or reasoning are possible ... The possibility of minds capable of forming progressively more objective concepts of reality (i.e. of decoding Nature) is not something the theory of natural selection can attempt to explain."*[35]

Nowhere is this inexplicable aspect of the human mind more

obvious than in the area of mathematics. Astrophysicist, Dr Paul Davies, Regents' Professor and Director of the Beyond Centre for Fundamental Concepts in Science at Arizona State University, writes:

> "The most striking product of the human mind is mathematics. This is a baffling thing. Mathematics is not something that you find lying around in your backyard. It's produced by the human mind. Yet if we ask where mathematics works best, it is in areas like particle physics and astrophysics, areas of fundamental science that are very, very far removed from everyday affairs. In fact they are at the opposite end of the spectrum of complexity from the human brain. In other words, we find that a product of the most complex system we know in nature, the human brain, finds a consonance with the underlying, simplest and most fundamental level, the basic building blocks that make up the world. That, I think, is an astounding and unexpected thing, and it suggests to me that consciousness and our ability to do mathematics is no mere accident, no trivial detail, no insignificant by-product of evolution that is piggy-backing on some other mundane property. It points to what I like to call the cosmic connection, the existence of a really deep relationship between minds that can do mathematics and the underlying laws of nature."[36]

In other words, the inexplicable ability of the human mind to engage in abstract thinking is made doubly inexplicable by the fact that the abstract concepts that we formulate accurately reflect the fundamental forces of nature and provide a meaningful explanation of them. Thus, not only is intelligent design obvious in the existence of the logical, orderly universal constants and fundamental forces of the universe, but also in our ability to decipher and quantify them. The hand of an intelligent Creator is evident in both the fundamental codes of the universe and in our ability to decode them. As theoretical physicist Dr Sir John Polkinghorne states, *"The reason within [matches] the reason without."*[37]

This is what Einstein was referring to when he remarked, *"the eternal mystery of the world is its comprehensibility"*.[38] Indeed, it is a great mystery for naturalists and evolutionists, because this profound intelligibility of the physical universe cannot be accounted for by natural selection. Our ability to construct abstract formulae to understand and explain the universe offers no logical benefit for our survival as individuals or as a species.

This particular argument for God's existence may seem, on the surface, a little esoteric, but those who follow it through to its logical conclusion find it extremely convincing. For instance, Roy Williams, author of *God Actually*, writes that this evidence of mankind's ability to engage in abstract thinking in order to understand the universe in his conversion to the Christian faith. He writes:

"It played an important part in my passage toward belief in Christianity ... because the theistic implications of this argument seemed to me inescapable, and still do."[39]

The inability of natural selection to provide an adequate explanation for the link between 'the reason within and the reason without' leads many, like Roy Williams, to believe in a supernatural Intelligent Designer. It is this link between our ability to engage in abstract thought and the discoverable laws of science that **require** abstract thought, that led theoretical physicist John Polkinghorne to conclude that God is the most logical explanation for these two phenomena. He wrote:

"There is some deeper rationality which is the ground of both, linking them together."[40]

For those with an open mind, the evidence for the existence of an intelligent designer is all around:

- The extraordinarily fine-tuned universal constants, which hold our universe together.
- Our miraculously shaped and strategically located planet that is perfectly designed for life.
- The unbelievably complex genetic information encoded into each of our cells.
- The mystery of human consciousness, in terms of both consciousness and cognition.

Evolutionists such as Richard Dawkins will, no doubt, continue to claim that *"there is not a tiny shred of evidence for the existence of any kind of god."*[41] In order to maintain this position in the face of such overwhelming evidence to the contrary, Dawkins and others are forced to adopt theories that can only be described as extreme and desperate. The most popular is the Multiverse Theory. This is the idea that an infinite number of universes exists and that, by sheer weight of numbers, the probability is that one of those universes will overcome the statistical improbabilities that we have discussed thus far and produce a life-permitting world.

Commenting on this imaginative theory, astrophysicist Paul Davies states;

> *"Invoking an infinite number of other universes just to explain the apparent contrivances of the one we see is pretty drastic, and in stark conflict with Occam's Razor (according to which science should prefer explanations with the least number of assumptions). I think it's much more satisfactory from a scientific point of view to try to understand why things are the way they are in this universe and not to invent imaginary universes to do the job."*[42]

In his book, *The Mind of God*, Davies also states;

> "My conclusion is that the many-universes theory can, at best, explain only a limited range of features, and then only if one appends some metaphysical assumptions that seem no less extravagant than design. In the end, Occam's Razor compels me to put my money on design."[43]

Davies is right to invoke Occam's Razor: the principle that if two or more explanations can account for all the facts, the simpler one is more likely to be correct. In this case, the simplest and most logical explanation for the abundant evidence of intelligent design within our universe is the existence of a supernatural Intelligent Designer. All other explanations require significantly more complex, imaginary and unlikely scenarios. This is what led astrophysicist, Dr Paul Davies, to conclude:

> "It may seem bizarre, but in my opinion, science offers a surer path to God than religion. People take it for granted that the physical world is both ordered and intelligible. The underlying order in nature - the laws of physics - are simply accepted as given, as brute facts. Nobody asks where they came from; at least they do not do so in polite company. However, even the most atheistic scientist accepts as an act of faith that the universe is not absurd, that there is a rational basis to physical existence manifested as law-like order in nature that is at least partly comprehensible to us. So science can proceed only if the scientist adopts an essentially theological worldview."[44]

Albert Einstein (1879 – 1955), perhaps the greatest scientist to have ever lived, and the physicist who developed the landmark Theory of General Relativity, regarded the obvious intelligent design of the physical universe as compelling evidence for the existence of God. He stated:

> "I want to know how God created this world. I am not interested in this or that phenomenon, in the spectrum of this or that element. I want to know His thoughts; the rest are mere details."[45]

Sir Isaac Newton said:

> "In the absence of any other proof, the thumb alone would convince me of God's existence."[46]

Of course, you don't need to be a scientist to discern the evidence of God's existence. It is clear and evident to anyone who has an open mind. The beauty and complexity of the universe is a symphony that declares the greatness of the God who created it. Thus, the Apostle Paul, in his letter to the Romans, writes:

> "What may be known about God is plain to them, because God has made it plain to them. For since the creation of the world God's invisible qualities—his eternal power and divine nature—have been clearly seen, being understood from what has been made, so that people are without excuse." (Romans 1:19-20).

6
EVIDENCE 4: THE EXISTENCE OF OBJECTIVE MORAL VALUES

In our examination of the evidence for God's existence, we now depart from the physical sciences and enter the realm of philosophy.

The existence of objective moral values is a very powerful piece of evidence for God's existence. The central thesis of this argument is that objective moral values CANNOT exist without God, and that the obvious existence of objective moral values is therefore strong evidence that God exists. This argument is comprised of two premises and a conclusion:

PREMISE 1: If God doesn't exist, objective moral values don't exist.

PREMISE 2: Objective moral values DO appear to exist.

CONCLUSION: Therefore, God exists.

In order to fully appreciate the logic of this argument, it is helpful to examine each of the premises separately.

. . .

PREMISE 1: IF GOD DOESN'T EXIST, OBJECTIVE MORAL VALUES DON'T EXIST.

This premise describes the concept that any moral standards which are absolute and objective are only possible if there exists some kind of ultimate, absolute standard against which all other standards are measured. If no such external standard exists, then human moral values are just one person's opinion versus another's.

Let me give you an illustration. Suppose you lived on an isolated island whose small community had never seen a tape measure or a ruler. The population had grown up with books that mentioned various units of measurement like inches and centimetres, but no one had ever seen anything to indicate the exact size of these units of measurement. People speculated as to how tall they all were, and often disagreed with each other about their relative heights, but without an objective means of accurately measuring height, they were only guessing. Then, one day, a tape measure washes up on the shore. Now they have an objective means of assessing their height. It is no longer a matter of debate or opinion. It is no longer one person's opinion versus another's. There is now an external, authoritative, objective ruler against which all former measurements and guesses can be compared. Measurement is no longer a matter of subjective opinion; it is now a matter of objective truth.

The same applies to moral values. Objective moral values can logically only exist if there is some absolute set of standards that exist APART FROM the subjective opinions of mankind, against which our various opinions can be measured and compared. Objective moral values are only possible if there is an external, transcendent set of absolute standards which can provide an inarguable, unchangeable 'measuring tape' for

Evidence 4: The Existence Of Objective Moral Values | 97

assessing human behaviour and attitudes. Without such an external, transcendent set of standards, morality becomes relative; a matter of personal opinion or, at best, group consensus.

The concept that moral standards are simply a matter of personal or group opinion is referred to as subjectivism or relativism. Indeed, relativism is the ultimate, logical extension of atheism. If God does not exist, then we can decide for ourselves what we will consider to be right and wrong. There is no one to tell us what to do, so we can make up our own rules and values.

This, of course, is precisely the direction in which our society has been heading during the last few centuries as we have gradually abandoned our previous Judeo-Christian roots and embraced secularism (a society divorced from religion). Relativism is now the predominant moral philosophy in our 'sophisticated' secular world: a philosophy which proclaims that there are no absolutes, that all values are valid and no one has the right to judge or criticise another person's values. In removing God from our foundation, we have thrown away our objective measuring tape and now, it appears, almost anything goes.

Some atheist will try to argue that it is still possible to have objective moral values without God. They will argue that things like murder and rape are objectively wrong and we don't need God in order to perceive that. But they are missing the point! I agree that murder and rape are objectively wrong, but only because horrible actions such as that go against the commandments and the very nature of our Holy God. But if atheists deny the existence of God, there is no longer any solid basis for objective morality, there is only the shifting sands of personal and public opinion.

Atheists will persist in their argument, claiming that objective standards can be ascertained via majority opinion or public consensus. But this is not a solid basis for *any* moral values, because the majority of people can be wrong. The majority of Nazis in Germany in the 1940s decided it was a good idea to exterminate millions of Jews and other minority groups. If objective values are determined by majority vote, then we are in big trouble, because the majority often get it horribly wrong!

Not only is it possible for the majority of people to get it wrong, but establishing societal values via majority vote will inevitably result in constant moral backflips, because society keeps changing its mind. Fifty years ago, the majority of people thought that homosexuality was indecent and a perversion, and homosexuals were even arrested. Today, society has changed its mind. The majority of people now believe that homosexuality is a beautiful expression of love, and we even allow homosexuals to marry. Where is the objective moral standard on this issue? Is our current view regarding homosexuality right, and the previous one wrong? Or *vice versa*? Both views were established via majority opinion, so which is the objective, absolute standard? They can't both be right.

You see, absolute, objective moral standards cannot be established via popular vote or public opinion. Objective moral standards require an external, transcendent set of standards that do not change with opinion or popularity.

But we haven't plumbed the depths of this issue yet. Let me go even further down a very dark path. If there is no God, then even murder and rape cannot be said to be wrong in an absolute, objective sense. The murderer or rapist is certainly going against the majority opinion by his actions; he is breaking a moral code that the vast majority have agreed upon. But on what basis can we objectively say that, even in these cases, the majority is right? The murderer or rapist can simply say, *"I disagree with your moral standards. You can't force your standards upon me.*" We can lock up the rapist or murderer and throw away the key, we can condemn him for breaking a popularly voted moral code, but without an absolute, external measuring tape there is really no logical basis for claiming that murder or rape are OBJECTIVELY wrong. It is just the murderer's or rapist's opinion against ours. Without a Holy, law-giving God, there is no ultimate accountability and no absolute moral code, and we are all left shuffling around in moral darkness.

Jeffrey Dahmer was one of the world's most infamous killers. He ate his victims and stored parts of them in his freezer. He was caught, convicted and imprisoned, where he was eventually murdered by another prisoner. Before his death, however, he was interviewed on the American TV program, *NBC Dateline*, and was asked why he did such terrible things. He stated,

> "*If a person doesn't think that there's a God to be accountable to, then what's the point of trying to modify your behaviour in order to keep within acceptable ranges. That's how I thought anyway. I always believed in the theory of evolution - that we all just came from slime and when we die, you know, that was it, there's nothing.*" [1]

Amazingly, before he was murdered in prison, Jeffrey Dahmer was genuinely converted to Christ. On the *Dateline* program he said,

> "*I have since come to believe that the Lord Jesus Christ is truly God and that I, along with everyone else, will one day be fully accountable to Him.*" [2]

What is particularly striking about Dahmer's confession is the correlation he made between his disbelief in God and the perceived absence of objective moral values. And in one sense, he was absolutely correct; if there is no God, there *are* no absolute, objective moral values. This is the ultimate extension of atheism, but very few atheists truly perceive this truth. The only way objective values can exist is if there is a God who has set in place standards which exist above and beyond mankind's shifting opinions.

Some atheists DO finally reach this conclusion. Richard Dawkins, the outspoken atheist and evolutionist, once admitted,

> *"Without God, there is no evil and no good: nothing but blind, pitiless indifference."*

Similarly, Fyodor Dostoyevski, the Russian novelist and philosopher had one of his characters declare:

> *"If there is no God, everything is permissible."*

These scholars have perceived a truth that few people fully grasp: without God, objective values simply cannot exist.

Before we finish with this first premise, let's be clear about something. I am not saying that atheists are incapable of living what we perceive to be comparatively good, moral lives. A person doesn't need to believe in God in order to be kind to others or to be honest and trustworthy and generous and selfless. It's not BELIEF in God that is essential for there to be objective moral values and for people to live by those values, it's the EXISTENCE of a law-giving God that is essential. If there is no God, then the values that people choose to live by, whether they be atheists or theists, have no objective basis; they are merely moral PREFERENCES rather than objective moral standards.

Evidence 4: The Existence Of Objective Moral Values | 101

Some atheists argue that objective moral goodness exists within the universe as a 'thing' – a kind of self-existent absolute standard which human beings can somehow sense. This concept is called atheistic moral Platonism, named after Plato who proposed something like this in the 4th century B.C. In other words, 'goodness' is a self-existent entity within the cosmos, totally independent of humankind's moral opinions. But this just doesn't make sense. How can the moral value of justice exist independently of people? As the philosopher Dr William Lane Craig states;

> "It's hard to make sense of this. It's easy to understand what it means to say that some person is just, but it's bewildering when someone says that in the absence of any people justice itself exists."[3]

Moral values are properties of conscious beings who interact with each other and the world, and they can't simply exist in their own right. In particular, OBJECTIVE moral values require the existence of a SUPREME being whose absolute values sit above the changing preferences and opinions of humankind.

Thus, the first premise is logically sound: if God doesn't exist, objective moral values don't exist.

PREMISE 2: OBJECTIVE MORAL VALUES DO EXIST

Here is the great dilemma for atheists: objective moral values DO exist. Even atheists admit this. In my ongoing debates and discussions with atheists I find that almost no one denies the existence of objective moral values in our world.

Almost all of us hold to a strong belief in absolute moral goodness and evil. For example, we recognise that child abuse, rape and torture are evil. They are not merely socially unacceptable because of some kind of majority vote. They aren't wrong simply because of some kind of communal subjective moral PREFERENCE. They are inherently evil actions, in and of themselves. There is something so IMPLICITLY evil about such things as child abuse and rape that we don't have to be told they are wrong; we simply KNOW it in the depths of our souls. We don't have to hold a national referendum to decide whether or not rape or child abuse are acceptable. The evil nature of such actions is so intrinsically obvious that we recoil from it with understandable disgust and those who commit such indecent acts are rightly regard as sick and evil.

In the same way, we don't need to hold a referendum to determine whether it is a good thing to love someone or to care for little babies. These are so obviously and implicitly good, in and of themselves, that they require no debate. Once again, those who err from these objective moral standards are considered to be morally sick.

C.S. Lewis, in his masterpiece, *Mere Christianity*, refers to the fact that mankind has an inbuilt sense of right and wrong, a sense of what is fair and decent and what is not, which he calls variously "the Law of Conscience", "the Law of Decent Behaviour" and the "Law of Human Nature" (not inferring that it has originated from blind nature, but simply that these values are "natural" to us).[4]

It is at this point that some people will claim that there is no such thing as a universal conscience or code of right and wrong, because different civilisations in various ages have held to different moralities.

But this is not true. There have been superficial differences between their moralities, but these have merely been differences in the application of deeper, more foundational moral principles. For example, societies have differed as to whether a man can have one wife or five, but all societies agree that a man cannot simply take any woman he likes at any time. Underlying the superficial variances in the application of morality there lies a deeper set of values that are universal. A study of the moral teachings of all cultures and societies of the past and present reveals an overwhelming agreement of fundamental values.[5]

To illustrate this universal agreement, you only need to try to imagine a society with a totally different set of values. C.S. Lewis explains:

> "Think of a country where people were admired for running away in battle, or where a man felt proud of double-crossing all the people who had been kindest to him. You might just as well try to imagine a country where two and two made five. Men have differed as regards what people you ought to be unselfish to—whether it was only your own family, or your fellow countrymen, or everyone. But they have always agreed that you ought not to always put yourself first."[6]

Lewis continues:

> "Although there are differences between the moral ideas of one time or country and those of another, the differences are not really very great—not nearly so great as most people imagine—and you can recognise the same law running through them all."[7]

Thus, while society may, at times, shift and change its collective opinion concerning the application of some moral values, there remains a core set of values that we regard as non-negotiable absolutes. Although our post-modern society claims to espouse relativism (the belief that there are no absolutes), in practice, the vast majority of the human race hold certain core values in place as absolutes. It is absolutely good to love and care for a baby, and it is absolutely evil to dash its brain out against a rock. It is absolutely good to love and care for a child and it is absolutely evil to torture and rape a child. As a society we are rightly incensed when we hear of clergy and priests who rape and sexually abuse children in their churches. Our moral outrage at such an action arises not merely from its departure from an agreed-upon moral preference, but from our sense, deep within our collective conscience, that the action itself is inherently, implicitly evil.

If you are completely honest with yourself, you will almost certainly agree that objective moral values really do exist. Our strong reaction when they are broken indicates the fundamental nature of those values that runs far deeper than mere collective opinion or referendum. These values are right or wrong, not because we voted on them, but simply because they just *are*.

C.S. Lewis comments:

> "We are forced to believe in a real Right and Wrong. People may be sometimes mistaken about them, just as people sometimes get their sums wrong; but they are not a matter of mere taste and opinion any more than the multiplication table."[8]

But this presents a major dilemma for atheism. Because if absolute values DO exist, and such values are only possible if God exists (as Premise 1 states), then the inescapable conclusion is that God must, therefore, exist. It is to avoid this conclusion that, very occasionally, an atheist will resolutely deny the existence of objective moral values. This is an extreme position and one which becomes increasingly bizarre if such a person is pushed to defend it.

William Lane Craig provides an example of such an extreme attitude. In his book, *"On Guard"*, he recounts being present at a forum on morals involving a panel of experts who made presentations and then answered questions from the audience. One of the atheistic panellists was adamant that objective moral values don't exist. After making her case for relativism, claiming that nothing is either good or bad in an absolute sense – that everything is just a matter of personal or communal preference - she opened the discussion for questions from the audience. I will let Dr Craig recount what happened in his own words:

> *"The next man who stood up said, 'Wait a minute. I'm rather confused. I'm a pastor and people are always coming up to me, asking if something they've done is wrong and if they need forgiveness. For example, isn't it always wrong to abuse a child?' I couldn't believe the panellist's response. She replied: 'What counts as abuse differs from society to society, so we can't really use the word abuse without tying it to a historical context.' 'Call it whatever you like,' the pastor insisted, 'but child abuse is damaging to children. Isn't it wrong to damage children?' And still she wouldn't admit it! This sort of hardness of heart ultimately backfires on the moral relativist and exposes in the minds of most people the bankruptcy of such a worldview."*[9]

Dr Craig is right. The moral relativist who rejects all moral absolutes is ultimately viewed with abhorrence by the rest of us, because there is something undeniably intrinsic about certain moral values. Our deep emotional response to such values indicates that they are more than mere collective opinions or preferences.

Some moral relativists will argue that our perceived objective moral standards are simply the product of social conditioning.

They claim that these moral preferences have arisen as a result of society instilling those values into our collective psyche – a form of mass indoctrination. Furthermore, they claim that the reason for society's preference for these values is not because of any inherent good or evil, but simply because those values are convenient for the advancement of society as a whole. According to this theory, we have come to uphold these values simply because they are convenient for our continuation as a species. For example, according to this view, moral precepts such as "don't hit people" and "don't lie" are ubiquitous not because they are rooted in some kind of pre-existing, transcendent, objective moral code, but simply because they have been arbitrarily instilled into us by society as a convenient means of maintaining order and optimising the quality of our lives. To put it even more simply, this view states that our moral values are TAUGHT rather than inherited from some transcendent source.

There are three things to say in response to this.

Firstly, just because something is taught to us, doesn't mean it is not objectively true. School children have to be taught mathematical concepts, but they are still objectively true. The statement that 2 + 2 = 4 is objectively true and the fact that it has to be taught to young children does not negate its objective truth.

Secondly, the moral instruction that takes place within society is not merely the imposition of arbitrary standards upon individuals – not some form of indoctrination - but rather, represents the REINFORCEMENT of pre-existing objective moral values that already reside within the conscience of the individual and within our collective conscience as a society. When parents teach children that it is wrong to hit someone, they are appealing to the pre-existing objective moral value that violence is wrong; that to inflict hurt upon another person is inherently wrong. Deep down everyone, even children, know this, but we need reinforcement, particularly at a young age, to train us to live in accordance with this value.

Thirdly, the strength of our emotional response to certain moral values indicates they are not mere convenient social preferences. Our abhorrence and revulsion in response to hideous acts of violence and abuse make no sense if the perpetrators are simply breaking convenient social conventions. When someone rapes a woman or sexually abuses a child a profound sense of outrage and disgust is generated from deep within us. We know implicitly, without having to be taught, that such reprehensible acts represent the breaking of an absolute, objective moral standard. We recognise such crimes as inherently evil and the perpetrators as sick, and this is reflected in our emotional response.

Dr Michael Ruse, a prominent atheist philosopher, admitted the existence of objective moral values, stating, *"The person who says it is morally acceptable to rape little children is just as mistaken as the person who says 2+2=5."*[10] He makes a good point. The formula, 2 + 2 = 5, is objectively wrong, and even if the whole world voted to say that they believe that 2 + 2 = 5, it would still be wrong, because objective truth is implicitly and independently true, irrespective of popular opinion. It's true, simply because it **IS**.

The same is true of objective moral values. They exist independently, apart from our opinions, and they have been stamped deep into our souls in such a way that they elicit a strong emotional response when they are broken or not adhered to.

C.S. Lewis states:

> *"Consequently, this Rule of Right and Wrong, or Law of Human Nature, or whatever you call it, must somehow or other be a real*

> thing—a thing that is really there, not made up by ourselves. ... It begins to look as if we shall have to admit that there is more than one kind of reality; that, in this particular case, there is something above and beyond the ordinary facts of men's behaviour, and yet quite definitely real—a real law, which none of us made, but which we find pressing on us."[11]

If both of the first two premises in the moral argument are true, then it logically follows that God must exist. If objective moral values are only possible if God exists, and if objective moral values DO, in fact, exist, then the unavoidable conclusion is that God exists. In this sense, it is a water-tight logical argument. To maintain atheism, one has to disprove one of the two premises of this argument, and it seems to me (and to many philosophers) that disproving either one is not possible.

Don't misunderstand me. The moral argument in this chapter and the scientific arguments in the preceding chapters are still a long way from revealing the God of Christian theology. But they do clearly point to a supernatural *Something*: an all-powerful, intelligent Creator of some kind. The next step is to see if there is any evidence for that *Something* being the God of the Christian Bible. This chapter has made the first tentative step in that direction because the existence of absolute moral laws speaks to the *character* of that *Something*.

The first three areas of evidence that we looked at in this book – the origin of the universe, the origin of life and the evidence of intelligent design – provide scientific evidence for the existence of a supernatural *Something*. But this fourth area of evidence – the existence of objective morality – provides strong evidence for the moral *goodness* of that *Something*. Indeed, the existence of an absolute moral code within the universe implies that the *Something* that created the universe is actually a *Someone* – a personal God who is morally wholesome and perfectly good, and who insists that we live according to his standards for the good of all.

Now, let us move on and see if there is any additional evidence that might point us more definitively to the God of the Christian Bible.

7

EVIDENCE 5: GOD'S INTERVENTION IN HUMAN HISTORY

Is there any evidence that the *Something* that created the universe, and which we have begun to suspect is a *Someone*, is actually the God of the Christian Bible?

We now move from the disciplines of science and philosophy to a study of history. Is there any evidence of the intervention of the Christian God in human history? After all, if God exists you would hope and expect that he hasn't just created the universe, set it in motion like a wind-up clock, then departed for an extended holiday, leaving us to our own devices. One would hope that the God who went to the trouble of creating the universe isn't some absentee landlord who doesn't care what goes on in his world after he has opened the front door and given us the keys.

So, is there any historical evidence of God's intervention in the world?

Yes! Absolutely! The Christian Bible is replete with accounts of God's regular intervention throughout human history, often involving extraordinary miracles. These range from miracles at the personal level such as healings of various diseases, to spectacular miracles at the national and global level; the parting of the red sea, the great flood, city walls miraculously collapsing, a river turning to blood, pillars of cloud and fire that led a nation through the desert, voices from thunderstorms, and dozens more examples. In fact, the Bible is positively brimming with accounts of God's extraordinary interventions in human history.

"But that doesn't count!" I hear you say. "The Bible is a religious book!" Of course it's a religious book. That's because it documents accounts of God's intervention in the world, which is EXACTLY what sceptics are asking for.

Can you see the circular contradiction that is operating here? (If I was writing this in the 1960s or 70s, I would refer to it as a 'Catch-22' scenario – but most people today wouldn't know what that means). Sceptics ask for historical, documented evidence of God's intervention in the world, but when it is presented to them, they dismiss it because it is written by people who have come to believe in God because of his intervention in the world!

The problem with these accounts in the Bible is that they are in the Bible. What I mean is that many people immediately dismiss the Bible as a fanciful fairy story written by religious fanatics and tend not to regard it as a reliable historical record.

In this chapter, I will firstly present some evidence as to why the Bible should be regarded as a reliable historical document. Secondly, I will then highlight a few incidents recorded for us in the Bible that provide credible, convincing evidence of God's existence.

THE BIBLE AS A RELIABLE HISTORICAL DOCUMENT

What I find utterly compelling about the Bible is that it is almost entirely written by people who started out not knowing God or not believing in him, but who became convinced of his existence and transformed by his presence in their lives. To the sceptic, I would say, *"the Bible is written by people who started out just like you, and that's what makes it so convincing."*

The first part of the Bible, the Old Testament, is the story of a nation (the ancient Israelites) whose founders and Patriarchs were called out of the ancient pagan world and gradually, over time, were taught to trust and follow God. It's not a very successful story, however, because each new generation of Israelites tended to drift away from God and even stop believing in him, needing to be convinced of God's reality all over again. It is a continual story of God repeatedly revealing himself in dramatic, sometimes miraculous ways, in order to convince a new generation of sceptics and apostates. It is this brutal honesty of the Old Testament that makes it so convincing. If you were making up a story about God, you would surely be much more likely to craft a story that depicted people who followed their God devotedly and were rewarded for their faithfulness.

The New Testament is even more impressive. Many of the writers of the New Testament wrote from a position of having once been a sceptic (particularly of the resurrection of Christ) but eventually became utterly convinced. The Apostle Paul, for example, started out arresting followers of Christ and putting them to death but was eventually converted when he encountered the risen Christ. Even the Apostles Matthew and John, who wrote two of the Gospel accounts of the life of Christ, were originally sceptical about the resurrection of Jesus and had to be convinced before they acknowledged its truth and came to believe in his divinity.

The Bible is a book largely written by sceptics who became convinced by God's miraculous intervention in our world.

But is there historical evidence to back up the many extraordinary claims about God's miracles that are recorded in the Bible? Are there other documents from antiquity or other historical evidence that substantiate the biblical narrative?

Yes, there are. But in the case of the Old Testament, they are scarce. That's because the narrative of the Old Testament spans a period of history from about 2500 B.C. to 430 B.C. There are very few literary works from that period of time that have survived through to the present day. Occasionally, however, historians and archaeologists uncover a document or an artefact that confirms a part of the Bible's story.

In his work, *Biblical Archaeology: Factual Evidence to Support the Historicity of the Bible,* Dr Paul L. Maier writes,

> "Ever since scientific archaeology started a century and a half ago, the consistent pattern has been this: the hard evidence from the ground has borne out the biblical record again and again — and again. The Bible has nothing to fear from the spade." [1]

Dr Maier goes on to cite many examples of archaeological discoveries which have verified the historical claims of the Bible. For example, up until the beginning of the 20th century, the Hittites were unknown outside of the Bible, and historians had long claimed that this was an example of the Bible's fanciful nature. However, in 1906, a Hittite city was uncovered east of Ankara, Turkey, along with clay tablets containing writing that provides detailed descriptions of Hittite culture.

There are literally dozens of similar examples – where critics claimed that something in the Bible was fictitious, only to have their criticism overturned by subsequent archaeological discoveries which validated the Biblical account. Here are some other examples of events, places or people mentioned in the Bible that were once doubted by historians, only to have these doubts subsequently overturned by archaeological verification:

- The Biblical cities of Haran, Hazor, Dan, Megiddo, Shechem, Samaria, Shiloh, Gezer, Gibeah, Beth Shemesh, Beth Shean, Beersheba and Lachish

- Shishak's Invasion of Judah (1 Kings 14)

- The Moabite Stone (2 Kings 3)

- Obelisk of Shalmaneser III (2 Kings 9–10)

- Burial Plaque of King Uzziah (2 Chronicles 26)

- Hezekiah's Siloam Tunnel Inscription (2 Kings 20)

- The Sennacherib Prism (2 Kings 18-19)

- The Cylinder of Cyrus the Great (2 Chron 36:23)

These Old Testament references to people, places or things which were once regarded as fictitious by sceptics, have been verified by archaeological digs or the discovery of corroborating historical documents.

A similar process of verification has occurred in regard to the New Testament accounts of the life of Jesus and the developing Christian church. Places frequented by Jesus, where he preached and supposedly worked miracles, have been uncovered by archaeological digs and verified by other historical documents.

More important to our discussion of God's possible existence, however, is the question of historical verification of the *miracles* recorded in the Bible. It is helpful to discuss the two major sections of the Bible separately in considering this question.

HISTORICAL EVIDENCE SUPPORTING BIBLICAL MIRACLES IN THE OLD TESTAMENT

The historical or archaeological evidence for ANY events in human history dating back between 2,500 to 4,500 years ago is extremely scant. The Bible is not alone in this. Time and the natural processes of weathering and decay assure that very few documents and artifacts have survived intact. In terms of verifying the miracle accounts in the Old Testament, we are faced with the additional problem that the nature of many of the reported miracles is such that there is no possible means of verification. When God spoke to people or manifested himself in the world in a physical way, such as the pillars of cloud and fire in the book of Exodus, these interventions would not have left any tangible evidence.

However, we are not left totally bereft of evidence of God's miraculous interventions. The miraculous collapse of the walls of Jericho is one such example. The Old Testament records that God called the Israelites to besiege the city of Jericho, as punishment for the particularly vile sins of its inhabitants. From the Bible and other historical records we know that this included human sacrifice, child sacrifice, witchcraft, entrenched prostitution, several vile sexual practices (see Deuteronomy 18:9-22; Genesis 15:15-16; Leviticus 18:24-30). The Israelites were commanded to march around the city once each day blowing their trumpets. On the seventh day, they were commanded to march around the city walls seven times and, after the seventh time, they were to give a loud shout and a blast on their trumpets and God would cause the city walls to collapse. And that, according to the Bible, is precisely what happened.

This, of course has been greeted with great scepticism by those outside of the Christian faith. However, archaeological excavations at the ancient site of Jericho during the latter part of the 20th century have added considerable weight to the biblical account. Early archaeological work in the late 1950s by a team led by Dame Kathleen Kenyon found the crumbled remains of heavily fortified double walls surrounding the ancient city. Dame Kenyon initially dated these walls as having been destroyed in about 1550 BC, roughly 150 years before the biblical Israelites supposedly besieged the city.[2] Her dating of the ruins led the sceptic world to conclude that the biblical account was, therefore, mythological. In the decades since Kenyon's dig, however, the remains of the walls have been dated using Carbon-14 dating as well as a more accurate examination of ceramic typology and glyphs found among the artefacts.[3] The result is that the walls are now dated at 1400 BC, which fits perfectly with the time the Israelites would have besieged Jericho. Dr Bryant G. Wood, in his paper, *"Did The Israelites Conquer Jericho?"* in *Biblical Archaeology Review (March -April 1990)* provided substantial evidence of the more accurate date of the walls and it is now almost universally acknowledged that Dame Kenyon misdated the ruins.

What was most important about Dame Kenyon's original findings, however, was the evidence that these ancient walls had collapsed *outward* rather than inward (as would normally be the case when a besieging army breached city walls). The walls were discovered to have consisted of a 3 to 5-metre wall of red bricks built on top of a 5-metre retaining wall (or revetment) of stone. Significantly, the collapsed upper wall of red bricks had not fallen inward onto the retained higher ground of the city, but had fallen outward and lay in ruins at the base of the retaining wall (revetment).

Dame Kenyon's notes make clear reference to this anomaly:

> *"... fallen red bricks piling nearly to the top of the revetment. These probably came from the wall on the summit of the bank ... the brickwork above the revetment."*[4]

This is extraordinary confirmation of the Hebrew text of Joshua 6:20 which literally says that when God caused the wall of Jericho to fall, the wall *"fell beneath itself"*.

Another recent discovery which appears to confirm the miraculous intervention of God in ancient times concerns the apparent identification of the biblical Mount Sinai. In the book of Exodus in the Bible, we are told that God led the Israelites to the foot of Mount Sinai to establish a covenant with them. We are then told:

> *"Mount Sinai was covered with smoke because the Lord <u>descended</u>*

on it in fire. The smoke billowed up from it like smoke from a furnace, and the whole mountain trembled violently" (Exod 19:18).

For many years the exact location of the biblical Mount Sinai was uncertain. A number of possible mountains were suggested, including Jabal al-Lawz and Mount Catherine, a peak near the Monastery of St Catherine in the Sinai Peninsula. But in 1984, the biblical Mount Sinai was identified by explorer Ron Wyatt as a mountain now known as Jabal Maqla. Although there still remains a degree of dispute, Jabal Maqla has been heralded by many as the biblical mountain because of several stunning features which align with the biblical account.

What is particularly fascinating about Jabal Maqla is that the entire top of the mountain is blackened, with rocks that have obviously melted, as if the summit has been scorched by some kind of intense fire. The rocks have a glassy black surface but when they are broken open, they reveal themselves to be red granite. Some kind of intense heat has been applied to the mountain top (which was never a volcano) that has literally melted the surface of the rocks. This extraordinary geological evidence accords with the Bible's description that God *"descended on it in fire"* and that the entire top of the mountain was like *"a furnace"*.

The same account in Exodus also describes how God commanded Moses to strike a rock with his staff. The rock split and water poured out, sufficient to quench the thirst of the entire nation which, by then would have numbered hundreds of thousands. Significantly, the same explorers who located Mount Sinai also discovered a huge split rock (16 storeys high) with ancient erosion channels and rounded, smooth rocks leading away from it for hundreds of meters towards a now-dry lakebed.

These discoveries, while not incontrovertible, do constitute an impressive degree of confirmation of some of the Old Testament accounts of God's miracle working power.

NEW TESTAMENT EVIDENCE OF GOD'S MIRACULOUS INTERVENTION

Evidence of God's hand at work in the world is nowhere clearer than in the life of Jesus Christ, as recorded in the New Testament. I will deal with the resurrection of Jesus separately in the next chapter, but at this point I want to draw your attention to the extraordinary accounts of his miracles. Because if these miracles are true, they represent extremely convincing evidence for the existence of God.

The four Gospels (accounts of the life of Jesus) in the New Testament describe the extraordinary life of Jesus Christ. I use the word 'extraordinary' deliberately. His life was extraordinary for several reasons:

- He claimed to be God in the flesh
- The Gospels record him performing many miracles
- The Gospels also record him rising from the dead three days after his cruel death by crucifixion

If these accounts are true, the 'God-question" has been answered unequivocally. But are they true? Or are they mere fables, fabricated by over-enthusiastic followers? Is there any evidence to support these Gospel stories?

THE FACT OF JESUS' EXISTENCE

I am occasionally amazed to meet someone who believes that

Evidence 5: God's Intervention In Human History | 121

Jesus never existed. This is an extreme viewpoint that completely overlooks the considerable weight of historical evidence supporting the life of Jesus. In my experience, the only people who hold this view are those who have no historical training and have not examined the historical evidence in any detail. The life of Christ is substantiated by a number of extra-biblical (independent) writers from antiquity, including Cornelius Tacitus, Lucian, Flavius Josephus, Suetonius, Pliny the Younger, Thallus, Phlegon of Tralles and Mara Bar-Serapion.

Respected historian, Dr Neil Carter, who is a professed atheist, writes:

> "I can't believe I'm feeling the need to do this, but today I'd like to write a brief defence of the historicity of Jesus. When people in the sceptic community argue that Jesus never existed, they are dismissing a large body of work for which they have insufficient appreciation, most often due to the fact that they themselves have never formally studied the subject.... The earliest writings which attest to the existence of Jesus come from the apostle Paul, a leather worker by day and preacher by night ... sometime in the mid-50s AD... The oral tradition which later came to inform the writing of the gospels predates the ministry of Paul by many years... Paul didn't invent these stories..."[5]

Dr Bart Ehrman is a respected historian who is a professed agnostic. He was recently interviewed by "The Atheist Guy" on "Atheist Radio" (an internet radio station whose sole aim is to discredit Christianity)[6]. Here is a transcript of part of that interview:

Atheist Guy: *Do you believe that Jesus actually existed?*

Dr Ehrman: *Yes. There is no serious historian who doubts the exis-*

tence of Jesus. There are a lot of people who want to write sensational books claiming that Jesus didn't exist, but I don't know any serious scholar who doubts the existence of Jesus.

Atheist Guy: *But there are historians who disagree with you, aren't there?*

Dr Ehrman: *None that I've ever heard of. Not serious historians. I know thousands of scholars of the ancient world and I don't know any one of these scholars who disagree.*

THE MIRACLES OF JESUS

Of course, it is one thing to concede that Jesus existed, it is quite another to accept the accounts of His miracles. And what were those miracles?

- On three separate occasions he raised dead people to life.
- On one of those occasions, Jesus stopped a funeral procession in the street and raised the dead person.
- On another of those occasions, the person he raised to life had been dead for three days.
- He instantly healed a wide range of diseases including leprosy and blindness.
- He instantly healed paraplegics and quadriplegics.
- He multiplied a small amount of food to feed thousands.
- He changed water into wine.
- He walked on water.

The extraordinary nature of Jesus' miracles has long been a point of contention for atheists. Having failed in their attempts to disprove Jesus' existence, the major thrust of the atheistic attack upon Jesus has been the attempt to discredit the veracity of these biblical accounts of his miracles. Many sceptics argue that Jesus' life was significantly embellished by the New Testament writers. According to this theory, Jesus was just an ordinary man who said some wise things and developed a popular following. After his death, his followers supposedly deified him (declared him to be God), concocting stories of supposed miracles and fabricating the myth of his resurrection.

This view was expressed and popularised by the liberal ecclesiastical scholar, John Spong, who claimed that the miracles recorded in the four biblical Gospels were *"a late-developing part of the Jesus story"*.[7]

INSUFFICIENT TIME FOR EMBELLISHMENT

This view is not supported by serious historians, however, who point out that there was insufficient time between the events of Jesus' life and their eventual recording in the gospels for embellishment to have taken place.

Professor A. N. Sherwin-White (1911-1993) was a world-renowned Greco-Roman historian who, among his many scholarly works, analysed ancient historical embellishments. He concluded that minor embellishment required a time gap of at least two generations, while major embellishment required at least 200 years.[8] In other words, even minor mythological embellishment can't gain traction if it is written within the lifespan of the first two generations after the event, because there are too many people still alive who know the facts and who could speak up and refute the embellishment.

Let me give you an example. At the time of this book's publication, it is 75 years since Winston Churchill led the Allies to victory over Adolf Hitler and the Nazis. Suppose I decided to write a biography of Churchill and I concocted all kinds of fanciful stories of him working miracles, raising people from the dead, walking on water and rising from the dead himself. I would very quickly be shut down as a fool and my written account would be overwhelmed by a flood of literary rebuttal. And in 2,000 years time, anyone investigating my ridiculous claims would uncover a veritable sea of indignant refutation. Even after more than 70 years have elapsed, I still could not get away with embellishing the life of Winston Churchill!

In the case of the life of Jesus Christ, the first three Gospels of Matthew, Mark and Luke were written only 30-35 years after Jesus' death, and the Gospel of John was written another 25 years later. In other words, the Gospels were written within the lifetime of the eye-witnesses to the events of Jesus' life. And the response of the ancient world to those Gospels was remarkable; the Gospel accounts were met with RESOUNDING SILENCE. No refutation. No protest on point of fact.

The reason for this is quite simple. No one could refute the Gospels because there were still thousands of eye-witnesses alive who had witnessed Jesus' miracles. Hundreds had witnessed Him raise people from the dead. Thousands had witnessed Him heal the sick, the blind, the paralysed. Thousands had witnessed Him multiply food in order to feed the hungry. As the Gospels were written and distributed in first century Palestine, there were literally thousands of eye-witnesses who could attest their veracity. That is why we do not find any literary refutation in the historical record. If the Gospel accounts were fanciful embellishments by some deluded followers, we would expect to find a flood of literary refutation. But there is none.

The complete lack of disputation and protestation in the first century in response to these extraordinary accounts of Jesus' miracles in the four gospels is particularly convincing. In fact, not only is there a complete lack of disputation, but some enemies of Jesus actually CONFIRMED his miracle-working ability.

EXTRA-BIBLICAL CONFIRMATION OF JESUS' MIRACLES

The Jewish sacred text, the Talmud, which was being written as Jesus was working these miracles, mentions Jesus on several occasions, even referring to his miracles. The Jewish rabbis could not deny His miracles, because they had been witnessed by, in some instances, thousands of people! Instead, they attributed them to the power of the devil and accused Jesus of *"sorcery"*. (It defies belief how anyone could possibly argue that the devil would be interested in healing the blind, healing the sick, raising up quadriplegics, feeding the hungry and raising people from the dead!) For example, The Babylonian Talmud (BT, Sanhedrin, 43a) states;

> *"On the eve of the Passover Yeshua [the Nazarene] was hanged. For forty days before the execution took place a herald went forth and cried, 'He is going forth to be stoned because he has practiced sorcery and enticed Israel to apostasy. Anyone who can say anything in his favour, let him come and plead on his behalf.' And since nothing was brought forward in his favour, he was hanged on the eve of Passover."*[9]

Significantly, even Jesus' enemies, the leading Jews of the day, could not deny the power of His miracles, instead choosing to attribute them to evil sorcery.[10] This is part of the reason why they eventually crucified Jesus. His claims to be divine (God) and his regular miracles were a major thorn in the side of the Jewish authorities, who considered him to be a blasphemer. Ultimately, they could neither silence him nor staunch the constant flow of miracles, so they executed him. The important point for our current discussion, however, is that even the Talmud, one of the Jewish sacred texts being written concurrently by Jesus' enemies, acknowledged his miracles. This is EXTREMELY strong historical verification.

The Jewish historian, Flavius Josephus, provides another source of verification when he described Jesus as *"a worker of wonders"*, which was a first century phrase for 'miracle worker'.[11]

Thus, Frank Morrison, in his pivotal book, *Who Moved the Stone?* states;

> *"The fact remains that the personal ascendancy and repute of Jesus during his own lifetime was immense. The stories of his cures of the blind, the paralytic and the possessed were widespread. They came from all parts of the country and were apparently implicitly accepted even in high quarters in Jerusalem."*[12]

The acknowledgment of Jesus' miracles by the Jewish authorities is also evident in their interaction with Jesus in an incident prior to his crucifixion. The Gospel writer, John, records a dramatic exchange between Jesus and the Jewish authorities in Jerusalem, shortly before his arrest. Jesus had arrived in Jerusalem to celebrate the Feast of Dedication and was preaching in the outer courts of the Temple. John writes:

> *"His Jewish opponents picked up stones to stone him [to death], but Jesus said to them, "I have shown you many miracles from the Father. For which of these do you stone me?" "We are not stoning you for your miracles, but for blasphemy, because you, a mere man, claim to be God." (John 10:31-33, NIV)*

The incident concludes with John's statement:

Evidence 5: God's Intervention In Human History | 127

"Again they tried to seize him, but he escaped their grasp." (John 10:39).

What is intriguing about this incident is the lack of denial by the Jewish authorities regarding Jesus' miracles. It would have been the perfect opportunity to deny the miracles of Jesus if they were mere fiction – stories concocted by his followers. The most logical explanation for the lack of denial by the Jewish authorities is that they simply *could not* refute the miracles because too many people had witnessed them. This confrontation with Jesus took place in a public courtyard with a large crowd of onlookers, many of whom would have already witnessed Jesus' miracle-working power. The authorities would have looked foolish denying them.

If we are looking for evidence of God's existence, the miracles recorded in the Bible and corroborated by external sources must surely rate as compelling. The miracles of Jesus, in particular, provide us with extremely convincing evidence of the hand of God at work in our world and clear corroboration of God's existence. The evidence for the historical veracity of Jesus' miracles is so strong that even the famous liberal scholar, Rudolf Bultmann, conceded:

> *"But there can be no doubt that Jesus did such deeds, which were, in his and his contemporaries' understanding, miracles, that is, deeds that were the result of supernatural, divine causality. Doubtless he healed the sick and cast out demons."* (Rudolf Bultmann, *Jesus*, Berlin: Deutsche Bibliothek, 1926, p. 159.)

If these biblical accounts of the miracles of Jesus are true, as I believe they are, then we have solid historical evidence confirming the existence of God.

Before we leave this topic, we must also briefly consider the implications of Jesus' miracles regarding his own claim of divinity. For Jesus did not merely wander around the countryside uttering platitudes and coining wise sayings. He was not merely a good moral teacher: he claimed to be much more. Throughout his brief ministry he made a series of extraordinary claims:

- He claimed to have existed eternally prior to his appearance on Earth (John 8:48-59; John 17:5)
- He claimed to be able to forgive sin (Matthew 9:5; Luke 7:48)
- He claimed that he would judge all mankind at the end of history (Matthew 25:31; Luke 18:8)
- He claimed to be God (John 8:58; John 10:30)

Of course, merely claiming something does not make it so. Psychiatric wards are full of people who believe they are all kinds of things. But there are two factors that cause us to sit up and take notice of Jesus' claims: his obvious sanity and his extraordinary miracles. In regard to the latter, one must ask *"What further actions would we expect God to perform if he manifested himself to us in human form?"* The miracles of Jesus and, as we shall see in the next chapter, his resurrection from the dead, force us to seriously consider his claim of divinity. We cannot simply dismiss his claims as the ravings of a mad man.

C.S. Lewis wrote:

> *"I am trying here to prevent anyone saying the really foolish thing that people often say about Him: 'I'm ready to accept Jesus as a great moral teacher, but I don't accept His claim to be God.' That is the one thing we must not say. A man who was merely a man and said the sort of things Jesus said would not be a great moral teacher. He would either be a lunatic—on a level with the man who says he*

is a poached egg—or else he would be the Devil of Hell. You must make your choice. Either this man was, and is, the Son of God: or else a madman or something worse. You can shut Him up for a fool, you can spit at Him and kill Him as a demon; or you can fall at His feet and call Him Lord and God. But let us not come with any patronising nonsense about His being a great human teacher. He has not left that open to us. He did not intend to."[13]

8

EVIDENCE 6: THE RESURRECTION OF JESUS CHRIST

Without a doubt, the resurrection of Jesus Christ from the dead is the most convincing evidence for God's existence. This is why it has always been the centre-point of the Christian faith. As the Apostle Paul wrote,

> "If Christ has not been raised, our preaching is useless and so is your faith." *(1 Corinthians 15:14)*

But let us be clear about something here. If Jesus did not rise from the dead, it would not prove that there is no God, because there is already such a compelling array of evidence. At worst, it would mean that Christianity is just another dead religion, trying to somehow connect with God but failing to do so.

If, on the other hand, Jesus really DID rise from the dead, it would be incontrovertible evidence substantiating both God's existence and the claims of Christ himself, who claimed to be none other than God in the flesh. For this reason, atheists have made a particular focus of trying to discredit the evidence for the resurrection. The problem they face, however, is that the historical evidence is impressive.

According to the Gospel accounts, Jesus rose to life and appeared to his followers over a period of 40 days, with Paul mentioning one occasion when he appeared to a crowd of over 500 people (1 Corinthians 15:1-11). The New Testament also records the inability of the Jewish authorities to produce his dead body in order to quash the news of his resurrection, as well as their failed attempt to spread the rumour that Jesus' body had been stolen by his disciples (Matthew 28:13).

Those are the bare facts, as reported by the biblical writers. So, what evidence is there to support this extraordinary story?

CORROBORATION BY HOSTILE WITNESSES

In a court of law, corroboration by hostile witnesses is extremely powerful. This refers to corroborating testimony by people who are NOT on your side, who may even be your enemies, but who confirm your story. In the case of Jesus' resurrection, there are two very clear instances of corroboration by sources outside the Bible, and both of these sources come from people who were NOT Jesus' followers.

Firstly, Phlegon of Tralles, a second century Roman historian, wrote of the supernatural darkness and earthquakes throughout parts of the Roman Empire at the time of Jesus' crucifixion:

> "There was the greatest eclipse of the sun and it became night in the sixth hour of the day so that stars even appeared in the heavens. There was a great earthquake in Bithynia, and many things were overturned in Nicaea." [1]

Even more extraordinary, was his corroboration of the resurrection itself:

> "Jesus, while alive, was of no assistance to himself, but that he arose after death, and exhibited the marks of his punishment, and showed how his hands had been pierced by nails."[2]

This is no small thing. Here we have a Roman citizen, not a follower of Christ and probably still entrenched in the worship of the Roman gods, who writes of Christ's resurrection as an established historical fact.

Then there is the famous reference by Flavius Josephus, a practising Jew and an outstanding first century historian. In 93 AD, in Rome, Josephus published his lauded work, *Antiquities of the Jews*, which included the following account:

> "About this time there lived Jesus, a wise man, if indeed one ought to call him a man. For he was one who performed surprising deeds and was a teacher of such people as accept the truth gladly. He won over many Jews and many of the Greeks. He was the Messiah. And when, upon the accusation of the principal men among us, Pilate had condemned him to a cross, those who had first come to love him did not cease. He appeared to them spending a third day restored to life, for the prophets of God had foretold these things and a thousand other marvels about him. And the tribe of the Christians, so called after him, has still to this day not disappeared."[3]

This passage has been embroiled in controversy since the 1700's, with atheists and sceptics arguing that the passage must be an interpolation by later Christian copyists, as a Jew would surely never have written such a favourable report of Christ. The ensuing debate throughout the subsequent centuries has generated the equivalent of a small library of academic papers and books, either defending or refuting the authenticity of this one small paragraph. In 1995, however, a discovery was published that brought important new evidence to the debate. It uncovered an earlier document that Josephus had used as a source document for his comments about Jesus, thus proving that they were not added by a later copyist.[4] While determined atheists still maintain that this paragraph must have been inserted by a Christian copyist, the vast majority of neutral academics are now of the opinion that the reference is largely authentic (with the possible exception of the words *"if indeed one ought to call him a man"* and *"He was the messiah"*, which are still disputed). This reference by Josephus represents a powerful corroboration by a hostile critic of the miracles of Jesus and the historicity of His resurrection.

THE MARTRYDOM OF THE APOSTLES AND MANY OTHER FOLLOWERS

A second category of extremely convincing historical evidence for Christ's resurrection is the willingness of the Apostles and many other first century Christians to die for their belief in Christ's resurrection. If these followers of Jesus knew that the resurrection was a lie, if their claims of having seen the resurrected Jesus were fake, surely, many of them would have recanted when faced with the prospect of death. After all, no one knowingly dies for a lie.

But die they did! In the years shortly after Christ's death and resurrection, many Christians were rounded up, arrested and threatened with death unless they recanted their belief in Jesus as the resurrected Son of God. This campaign of terror was conducted at first by the Jewish authorities and then, a little later, was taken up by the Roman emperor. These events are recorded not just in the Bible, but in other historical documents from the first and second centuries. What is truly remarkable is the number of Christians who refused to recant. Many of them were adamant that they had actually seen the resurrected Jesus and they refused to renounce their faith in him, preferring to die rather than deny his resurrection.

Of the twelve Apostles appointed by Jesus, eleven were ultimately put to death for refusing to cease proclaiming their message about the resurrection and divinity of Jesus. Many hundreds of other first century Christians were also put to death at the hands of both the Jews and the Romans, because of their unwillingness to renounce this belief.

Nobody is willing to die for something they _know_ to be a lie; at least no one who is sane! If the resurrection was not true, if it was a fabrication invented by the disciples, they would not have been willing to die for it. The fact that the Apostles and many other first century Christians went willingly to their deaths at the hands of the governing authorities, refusing to renounce their belief in Christ's resurrection, is extremely compelling historical evidence for its authenticity.

"OK guys - here's the plan! We get the body out of the tomb and stash it somewhere. Then we come back here and tell a story that will probably get us all killed. So who's with me on this?"

Dr Luke Johnson, a New Testament scholar from Emory University, argues that only something as significant as the resurrection of Jesus can account for the dramatic transformation of the first century disciples from timid followers to bold evangelists who were willing to die for their beliefs. He states:

> "Some sort of powerful, transformative experience is required to generate the sort of movement earliest Christianity was...."[5]

Similarly, N. T. Wright, an eminent British scholar, concludes:

> "That is why, as a historian, I cannot explain the rise of early Christianity unless Jesus rose again, leaving an empty tomb behind him."[6]

SIX UNASSAILABLE FACTS IN THE RESURRECTION NARRATIVE

The historical narrative of Christ's resurrection on that first Easter Sunday morning, contains six unassailable facts that are accepted by even the most ardent atheist historians and which provide strong support for the veracity of the resurrection.

FACT 1: THE PUBLIC BURIAL OF JESUS IN THE TOMB OF JOSEPH OF ARIMATHEA.

This is extremely significant, because it means that the location of the tomb was known to the people of Jesus' day. The fact that it was publicly known as Jesus' burial place means that any false claims of his resurrection could easily have been thwarted simply by rolling back the tombstone. Dr John A.T. Robinson, from Cambridge University stated that the burial of Jesus in a well-known tomb is one of *"the earliest and best attested historical facts about Jesus."* [7]

FACT 2: JESUS' TOMB WAS DISCOVERED TO BE EMPTY ON SUNDAY MORNING

Historians, even those who are atheists and sceptics, have corroborated the fact of the empty tomb. Dr Jakob Kremer, an Austrian historian, states,

> *"By far most scholars hold firmly to the reliability of the biblical statements concerning the empty tomb."* [8]

Similarly, Dr D. H. van Daalen states,

> *"It is extremely difficult to object to the empty tomb on historical grounds; those who deny it do so on the basis of theological or philosophical assumptions."* [9]

FACT 3: MANY PEOPLE WITNESSED POST-RESURRECTION APPEARANCES OF JESUS

The fact that many of Jesus' followers had post-resurrection experiences of him is conceded by most serious historians today. Thus, even Dr Gerd Lüdemann, a prominent sceptic, admits:

> *"It may be taken as historically certain that Peter and the disciples had experiences after Jesus's death in which Jesus appeared to them as the risen Christ."* [10]

FACT 4: THE AUTHORITIES WERE NEVER ABLE TO PRODUCE JESUS' BODY

Significantly, and I mean VERY SIGNIFICANTLY, the first century Jewish and Roman authorities were never able to produce the dead body of Jesus in order to quell the rumours of his resurrection. It would have been the obvious trump card! All they had to do was drag out his dead body and say, *"No, you foolish Christians, he isn't risen! Here is his rotting corpse! Now go away and stop bothering us!"* But they didn't. They couldn't. Because the body was no longer there.

FACT 5: THE POSTING OF A ROMAN GUARD

The inability of the Jewish and Roman authorities to produce the dead body of Jesus is made even more significant by the extreme measures that those authorities undertook to ensure that Jesus could not be seen to rise from the dead in the first place. Matthew's Gospel records that the authorities had been aware of Jesus' previous claims that he would rise from the dead, so they had posted a Roman guard at the tomb. This was not to stop Jesus rising from the dead (because they did not believe he would) but to stop the disciples from stealing the body in order to perpetrate a hoax.

Now, let's be really clear about this: the posting of a Roman guard was a HUGE thing! The typical Roman guard was a sixteen-man unit that was governed by very strict rules. Each member was responsible for six square feet of space. The guard members could not sit down or lean against anything while they were on duty. If a guard member fell asleep, he was beaten and then burned to death. But he was not the only one executed: the entire sixteen-man team was executed if any one of the members fell asleep while on duty! This guard was a SERIOUS security step by the authorities to prevent any hint of a resurrection.

Matthew records the fact that after the resurrection, the Roman guard reported back to the Jewish authorities who then bribed them to spread the rumour that the disciples had stolen the body (Matthew 28:11-15). Dr Bill White, makes several critical observations regarding the fact that the Roman guard were NOT put to death for their failure to guard the tomb adequately:

> "If the stone were simply rolled to one side of the tomb, as would be necessary to enter it, then the authorities might be justified in accusing the men of sleeping at their posts, and in punishing them severely. If the men protested that the earthquake broke the seal and that the stone rolled back under vibration, they would still be liable to punishment for behaviour which might be labelled cowardice. But these possibilities do not meet the case. There was some undeniable evidence which made it impossible for the chief priests to bring any charges against the guard. The Jewish authorities must have visited the scene, examined the stone, and recognized its position as making it humanly impossible for their men to have permitted its removal. No twist of human ingenuity could provide an adequate answer or a scapegoat and so they were forced to bribe the guard and seek to hush things." [11]

Both the placement of the Roman guard at the tomb and the subsequent lack of punishment of the members of the guard, are strong evidence that the body of Jesus was not stolen by his disciples. It is completely unthinkable that a scattered band of disheartened disciples would somehow be able to steal Jesus' body out from under the noses of a professionally trained Roman guard who were commissioned with guarding the tomb WITH THEIR VERY LIVES! Any theories regarding body-snatching simply have no historical appreciation of the serious nature of this guard unit.

FACT 6: THE PLACING OF A ROMAN SEAL ON THE TOMB

Furthermore, Matthew's Gospel indicates that not only was a huge stone rolled across the entrance, but the tomb was then sealed with a Roman seal. This is acknowledged by most historians as a reliable historical record at this point. The seal referred to by Matthew was the seal of the emperor himself. It meant that anyone who broke the seal and opened the tomb would be put to death. The Romans did everything possible to ensure that the dead body of Jesus would stay safely in the tomb!

Given these six historical facts, it is inconceivable that a bunch of frightened, bedraggled disciples could have overpowered SIXTEEN trained, lethal, hardened Roman soldiers, carried off the body and perpetrated a myth.

THE BIBLICAL RECORD

With those six facts in mind, let me present a brief summary of the events surrounding the resurrection of Jesus as recorded by the New Testament writers.

Matthew's Gospel describes an earthquake that shook the tomb and nearby precincts on Easter Sunday morning, causing the stone to roll away, revealing an ALREADY EMPTY TOMB. It also records that a supernatural apparition appeared as the stone was rolled away, accompanied by a thunderous voice, seemingly from heaven, and that the entire Roman guard were so terrified that they fell to the ground (Matthew 28:1-4).

The New Testament goes on to record numerous subsequent public appearances by the risen Jesus, over a period of 40 days, sometimes involving crowds of up to 500 people at once (1 Corinthians 15:1-8). Even the Jewish writer, Flavius Josephus, makes a reference to the post-resurrection appearances of Jesus, in his book, *The Antiquities of the Jews*.

Significantly, the Gospels indicate that the disciples had no expectation of Jesus' resurrection. In his Gospel, Luke records the reaction of the disciples when Jesus first appeared to them after his resurrection:

> "They were startled and frightened, thinking they saw a ghost. He [Jesus] said to them, 'Why are you troubled, and why do doubts rise in your minds? Look at my hands and my feet. It is I myself! Touch me and see; a ghost does not have flesh and bones as you see I have.'" (Luke 24:37-39).

The recorded incredulity of the disciples at seeing the risen Jesus seems to refute the theory that they had mounted a commando-type raid on the tomb to rescue his body and perpetrate a hoax.

THE PROBLEM FOR SCEPTICS

If sceptics are going to maintain that Jesus did not rise from the dead, they have to reject not only the substantial evidence surrounding these New Testament accounts, but also common sense itself. Where did the body go? Why were the first century authorities unable to quash the rumours of Jesus' resurrection by producing his body? Why did they remain silent? The explosively expanding movement of Jesus' followers that resulted from his resurrection was surely a greater thorn in the side of Jewish leaders than Jesus had been, individually, prior to his crucifixion. The burgeoning Christian movement quickly resulted in a major split from Judaism and a significant loss of power for the Jewish authorities. Their desire to stamp out this new movement is evident by the extreme measures they undertook in arresting many of Jesus' followers and putting them to death. Given those extreme measures, it doesn't make sense that the authorities refrained from producing the dead body of Jesus – which would have been the ultimate and most effective measure for disproving the claims of this new 'religion'.

The most logical conclusion is that there was no dead body for the authorities to produce, because Jesus had, indeed, risen from the dead. All the evidence points to the fact that the body of Jesus WAS snatched from the tomb, but not by people. It was snatched from death and restored to life by God himself.

If the resurrection of Jesus is true, then the 'God question' has been answered unequivocally.

ATHEIST HISTORIANS CONCEDE THE FACTS

Let me point out that most of the historians quoted in this chapter are not Christians. They are respected historians, many of whom who are religious sceptics, but their study of the facts leads them to concede the veracity of at least some of the extraordinary evidence pointing to the resurrection of Jesus. Lay people who are untrained in history and who attempt to dismiss the resurrection by claiming that Jesus may not have actually died on the cross, but merely swooned and subsequently revived when placed in the tomb, don't find support among those who have examined the historical evidence. For instance, Joachim Kahl, an outspoken critic of Christianity, once grudgingly wrote that the crucifixion and death of Jesus was *"a probable fact"* and referred his readers to the extensive list of evidence.[12]

In his paper, *"The Evidence for Jesus"* [13], Christian apologist Dr William Lane Craig recounts a debate on the resurrection of Jesus that he had with a professor at the University of California who was a religious sceptic. The professor in question had written his doctoral dissertation on the resurrection narratives and was, therefore, thoroughly familiar with the historical evidence. During the debate, the sceptical professor conceded all the facts that I have briefly mentioned above, including burial in a publicly known tomb, discovery of the verifiably empty tomb, numerous post-resurrection appearances of Jesus to his followers, and the disciples' vehement post-resurrection beliefs and martyrdom. He could not deny these historical facts.

So, how did he explain them in order to maintain his atheism? His answer: he theorised that Jesus must have had a secret identical twin brother who stole his body from the tomb and masqueraded as the risen Jesus thereafter! This ludicrous theory is, of course, a work of pure imagination, because it has absolutely no historical evidence to support it. But it is helpful because it demonstrates two things:

- It demonstrates the extraordinary lengths to which sceptics will go in order to deny the resurrection and maintain their atheism.
- It also demonstrates the fact that historians who have actually studied these events at depth CANNOT DENY the historical facts that provide such strong evidence for the resurrection of Jesus. The key corroborating facts are beyond dispute.

SCEPTICS CONVERTED BY INVESTIGATING THE RESURRECTION

If the accounts of the resurrection of Jesus are true, we have incontrovertible evidence of the existence of God. Realising this, several academics and historians have, over the years, set out to disprove the resurrection story. Their philosophy was simple; disprove the resurrection and you remove the strongest argument for the existence of God.

Not only have these attempts been unsuccessful, they have regularly led to the complete capitulation and conversion of those undertaking the research:

Sir William Mitchell Ramsay (1851-1939) was a highly respected historian and archaeologist from Scotland. He set out to prove the historical inaccuracies of Luke and Acts. He spent 15 years researching and digging, only to end up being convinced of the incredible accuracy of the New Testament. He converted to Christianity and called Luke (the writer of two books of the New Testament) one of the greatest historians to ever live. He wrote several books on the subject, which have stood the test of time. His work caused an outcry from atheists because they had been funding his research and were eagerly awaiting his results in discrediting the validity of the New Testament![14]

Albert Henry Ross (1881-1950) was an English journalist and author who set out to disprove the "myth" of the resurrection. He was planning on writing a paper called "*Jesus – The Last Phase*", but he became converted during the course of his investigations. He ended up writing the classic book *Who Moved The Stone?*[15] under the pseudonym Frank Morrison. The book has led many people to faith in Christ.

Lee Strobel (born 1952) was a journalist for the Chicago Tribune. His wife converted to Christianity and Lee became very concerned. In order to "rescue" his wife from the church, he set out to disprove Christianity, focussing on the resurrection story. He spent 18 months, utilising his skills as a researcher and investigative reporter, interviewing experts from around the world, and studying the 1st century documents for himself. The overwhelming evidence for the resurrection of

Christ eventually led to his own conversion, and he went on to write the now famous book, The Case For Christ[16] (which was also made into a movie), along with several other books in the series, which have led many people to faith in Christ.

Simon Greenleaf (1783 - 1853) was a professor of law at Harvard University and a towering figure in the legal world. He set out to disprove the Gospel accounts, particularly the narratives concerning the resurrection of Jesus. He rigorously applied the criteria of historical reliability to the Gospels to test their authenticity, and applied cross-examination legal techniques to assess the reliability of the eye-witness accounts. Eventually he, too, was converted to Christianity, and his rigorous approach was foundational in the development of modern juridical Christian apologetics.[17]

C. S. Lewis, (1898 - 1963) was a distinguished professor of English literature at both Oxford and Cambridge universities (at different times) and is widely regarded as one of the greatest minds of the 20th century. He was originally an ardent atheist who fought long and hard to maintain his disbelief. In his book, Surprised by Joy, he wrote of his eventual conversion, stating, *"In the Trinity Term of 1929 I gave in, and admitted that God was God, and knelt and prayed: perhaps, that night, the most dejected and reluctant convert in all England."*[18] Lewis's inability to refute the solid historical evidence for the life and resurrection of Jesus led to his reluctant conversion – a conversion that would eventually result in the publication of, arguably, the most profound Christian philosophical writings of the modern era.

It is the reluctance of these learned scholars to believe the evidence, that makes their eventual conversions all the more significant. Indeed, history is replete with respected scholars who set out to disprove the resurrection of Jesus and were eventually convinced by the overwhelming weight of historical evidence.

The miraculous life and ultimate resurrection of Jesus Christ is the strongest possible evidence for the existence of God, recording his extraordinary appearance on our planet over 2,000 years ago.

A PERSONAL CHALLENGE

What about you? Are you one of those who has conveniently dismissed the resurrection of Jesus as a religious myth without really investigating the facts for yourself? Have you written off the Easter story simply because you find it too hard to believe? After all, miracles like that are impossible, aren't they?

No, they're not. Not if an all-powerful God exists. If there is a God who *spoke into existence* the whole universe with its untold billions of galaxies, who called it forth from nothing by the power of his command, then, surely, he can easily breathe new life into a single dead body. And if the resurrection of Jesus is true, then we must seriously consider the claim that Jesus repeatedly made that God will one day raise *all of us* from the dead too, to stand before him and give an account of our lives.

You see, if the resurrection of Jesus is true, then it changes everything! It means that there *is* a God, and one day, on the other side of your own impending death, you will meet him face to face.

Only those who have not taken the time to study the historical evidence dismiss the resurrection narratives in the Gospels lightly.

9

EVIDENCE 7: PERSONAL EXPERIENCE

If we are evaluating evidence for the existence of God, the personal experience of millions of Christians today must also be seriously considered. Many Christians testify that they have encountered God in a profound and personal way. This is in accord with God's promise in the Bible:

> "You will seek me and find me when you seek me with all your heart. I will be found by you, declares the Lord." (Jeremiah 29:13-14)

God promises to interact with us personally if we approach him with a humble, sincere heart. This has certainly been my own experience. I experience God's presence in my life on a daily basis in subtle ways, and, occasionally, in very profound ways.

What is to be made of this kind of experiential claim by millions of Christians? Obviously, just because people claim to experience something doesn't necessarily mean that their experience reflects reality. People can easily be self-deluded. The fact that large numbers of people attest to experiencing God's presence does not prove the validity of their experience, because large numbers of people can be wrong! There are people who are convinced that they have experienced all kinds of things; abduction by aliens, communication with aliens, the magical power of crystals, communication with the dead, the ability to leave their bodies and astral-travel, and much more.

On the other hand, personal testimony is considered to be a valuable component of legal proof when it is substantiated by other verifiable evidence. In a court of law, the testimony of just two corroborating eyewitnesses is often sufficient to obtain a clear verdict.

The testimony of Christians regarding their encounter with God must be considered seriously on a number of grounds, incorporating the same criteria that are used to evaluate any testimony in a court of law:

- **Volume:** Not just hundreds, nor thousands, but millions of Christians today and throughout the ages testify to experiencing God in a profound way in their lives.
- **Corroboration:** There is overwhelming corroboration in the testimony of Christians regarding their experience of the presence of God in their lives. Their experience of God's protection, guidance, peace and joy, together with countless testimonies of answered prayer and miraculous interventions, is a particularly impressive homogenous body of evidence.
- **The Character and Integrity of the Witnesses:** In a court of law, the character of the eyewitnesses will determine the weight that is given to their testimony. Unstable "crack pots" are not usually even given a hearing, whereas the testimony of sane, intelligent, coherent eyewitnesses is accorded great weight.

Among the Christian community who claim to experience the presence of God in their lives, there are highly intelligent, sane, rational, clear thinking people, not given to wild imaginative delusions. They are drawn from all walks of life, from straight-shooting manual labourers to highly qualified scientists, doctors, lawyers and other professionals.

While the testimony of millions of Christians who claim to experience God personally may not be acceptable to the sceptic as definitive proof of God's existence, it cannot not be lightly dismissed.

THE EVIDENCE OF MIRACLES

Many people claim to have either witnessed a miracle or been the actual subject of a miracle themselves. Once again, I concede that just because someone claims to have experienced something doesn't make it true. The human propensity for self-delusion can't be over-estimated. People can wish something to be true so strongly that they convince themselves that it is so.

I remember an encounter I once had with an extremely zealous Pentecostal Christian who believed that God has promised every Christian that they no longer have to accept sickness and disease. He was convinced that healing has been promised to every Christian for every disease and sickness. One Sunday morning, as I chatted with him at the door of the church where I had just preached, he proudly declared to me that God had cured him of his cold this week. Strangely, as he said this, his nose was running, his eyes were watering, his sinus cavities were completely congested and he was coughing. He looked terrible! When I pointed this out to him (aren't I helpful?), he declared that this was merely the devil trying to fool him and convince him that he wasn't healed. I saw the man again on the following Sunday and he still manifested all the same symptoms. If anything, he looked even worse! When I asked him how his healing was going, he once again declared his belief that he had been healed for over a week and that these symptoms were just the devil trying to undermine his faith.

I share this story with you because I sympathise with those of you who are sceptical of miracle stories. I am often sceptical too because I know how easily religious fervour can generate self-deluded miracles.

But not everyone who claims to have experienced God's miraculous intervention can easily be dismissed as a self-deluded religious zealot. When investigating claims of miracles, we must be careful not to 'throw the baby out with the bathwater' (what a horrible saying!).

Let me recount an experience I once had of God's miraculous power.

It was Christmas holidays, 1985. I was about to commence my fourth and final year of theological training to complete my Bachelor of Theology. My wife and I were on holidays on the Central Coast of NSW and it was the last day before we headed back home for me to recommence my studies. It was a beautiful sunny day and we went to the beach for the morning. The northern end of the beach, where the lifeguards were patrolling, was quite crowded with swimmers, body surfers and board riders all vying for a share of the waves, but there was literally no one at the southern end of the beach, about a kilometre away. Not wanting to contest the waves with a zillion other people, I left my wife at the patrolled end of the beach and jogged down to the southern end with my board. I spent about an hour in the surf, enjoying a really nice surf break which I literally had all to myself.

Then disaster struck. I was riding a particularly large, powerful wave when it sucked up on a sand bank and I wiped out badly. In the churning maelstrom, my board smashed violently into my leg and I knew I was in trouble. The pain was excruciating, shooting up from my leg and seeming to envelop my entire body. I was instantly sure I had broken my leg. I was in so much pain I couldn't move and couldn't get back onto my board. All I could do was cling onto the board and let the whitewash of the next few waves gradually wash me into shore. When I finally reached the shallows, I managed to reach the dry sand by shuffling myself backwards out of the water while sitting on my bottom and trailing my legs behind me.

When I finally reached dry sand, I looked down at my leg. There was a huge purple swelling on my shin where the board had smashed into me and the leg looked strangely distorted. The pain was intense and, having broken a number of other bones over the years (sporting injuries!), I was sure that my leg was broken. I was also sure there was no way I was going to be able to get back to my wife and get help on my own, because I couldn't bear to even put my foot to the ground. I was completely isolated at the southern end of the beach: the nearest people were nearly a kilometre away. I knew I couldn't possibly hop for a kilometre while carrying a surfboard, so I decided to sit on the sand and hope that someone would eventually come along to help me.

At first, I was angry and incredibly frustrated because I knew I was facing many weeks with my leg in plaster. I wouldn't be able to run (as I did every day – and still do). I wouldn't be able to ride my bike. And I would have lots of trouble at College because all of my lecture rooms were on the first floor (one floor up from the ground floor) and there were no lifts in the building.

After a while I began to feel convicted about my anger and frustration. I realised that it was pointless being angry about something that I couldn't change and that would only be a temporary nuisance. I also realised that I had very little to complain about compared to many people in the world who endure terrible hardship every day of their lives. I asked God to forgive me for my attitude and I began to think of all the blessings I should be thanking God for, instead of whingeing about a broken bone that would soon be fixed. As I reflected on these things, a simple worship song came to mind and there, on that lonely stretch of beach, I started to sing it.

I ended up singing several worship songs and after a while it suddenly dawned on me that my leg wasn't hurting anymore. I looked down at it. The huge, ugly purple swelling was completely gone. Maybe I was looking at the wrong leg? I glanced at the other leg. No. I was definitely looking at the correct leg. I looked back at the 'injured' leg. There was absolutely no sign of any injury. I poked and prodded the area where the injury had been. No pain! Feeling absolutely stunned, I stood up, standing only on my uninjured leg and holding my injured leg off the ground. Then I tentatively placed that foot to the ground, ever so gently. No pain! I placed all my weight on my injured leg. No pain! I started jumping up and down on my injured leg, trying to find any vestige of pain. None!

Still feeling completely stunned, I picked up my board and ran all the way back to my wife. When I arrived there, she was sunbathing on her towel, having just come out of the water. I stood over her, panting after my run, trying to think of how to explain what had just happened to me. In what I would later come to regard as the understatement of a lifetime, I simply said, "*I think I've just been healed!*"

Now, you don't know me. For all you know, I could be just another religious nutter with a self-deluded story of a miracle. After all, there are a lot of nutters out there!

The thing about self-delusion, however, is that it usually occurs when there is a pre-existing expectation of something. The man I previously described to you, who thought he was 'healed' of a cold, was fully EXPECTING to be healed and passionately WANTED to be healed, and so he convinced himself that he WAS healed. That's how self-delusion usually works.

In my case, however, I had NO expectation of being healed. In fact, quite the contrary. My healing took me completely by surprise. So much so, that it took me a while to be convinced it had actually happened. It took several minutes of poking and prodding and jumping before I finally became convinced that something miraculous had occurred. (If someone had been able to see me jumping up and down on the sand as I was testing my leg, they would have thought I had 'lost the plot'!)

This is certainly my most 'spectacular' encounter with God's miraculous power, but it is not the only one. I experience God's intervention in my life on a regular basis, often in much more mundane ways. I regularly experience his peace in times of difficulty and his guidance in times of uncertainty. Allow me to briefly give you some notable examples.

In the mid 1990's I was the minister of a church in Sydney, Australia, when I received an approach from another church to come and be their Senior Minister (these days they call them 'Lead Pastors'). My wife and I were considering the invitation and were praying daily to know God's will in the matter. We eventually reached a point where we felt that God might be leading us to accept the invitation. One morning I decided to phone the current Senior Minister of the church in question to ask him a few more questions. Before I phoned, I prayed and asked God for a clear, definitive sign one way or another. Immediately after finishing that prayer, I picked up the phone and dialled the Senior Minister.

That's when something strange happened. After dialling the number, there was no 'ringing' sound on the other end of the line, just a 'click' followed immediately by the hum of an open line. A moment later there was the sound of someone trying to dial a number. Thinking that my wife had picked up a phone in another room of our house and was trying to dial out, I called out, *"I'm using the phone!"*. Then I heard a male voice on the other end of the line. The strange conversation that followed went like this:

VOICE: "Hello? Who's this?"

ME: "It's Kevin Simington. Who's this?"

VOICE: "It's Dave." (The Senior Minster I was calling).

ME: "Oh! I didn't hear the phone ring on your end of the line."

DAVE: "It didn't ring. I phoned you."

ME: "No. I just phoned you."

DAVE: "What? No. I phoned you."

We eventually worked out what must have happened. We had both picked up our phones to dial the other person at almost precisely the same time. My call to Dave must have been slightly ahead of his and he had apparently picked up his phone to make his call in the split second that the connection from my phone had been made but before his phone had time to ring. What is even more extraordinary is that he, like me, had just been praying about his church's invitation to me and had asked God for confirmation. He had then felt prompted to call me.

The probability of two people phoning each other without prearrangement at exactly the same time and CONNECTING without either of their phones ringing must be infinitesimally small. When you also add the fact that Dave and I had both independently just prayed for God to give us a clear confirmation of my proposed move, the probability of all of those circumstances coming together by sheer chance moves into the realm of the unbelievable.

Needless to say, I accepted the call to that church.

Let me tell you of another incident. In 2018, after publishing three books on Theology and Philosophy, I was mid-way through writing my first science fiction novel. I was enjoying writing the novel and I felt that it was a good story, particularly as it portrayed a Christian worldview while still offering readers a cracking sci fi yarn. But part-way through the novel I began to have doubts. Was writing a sci fi novel too frivolous? Did God want me to be writing more serious books? Was I being self-indulgent? I became so plagued with doubts that I stopped writing. I prayed about the issue for several days, but no bolt from the blue hit me. Then, one afternoon, I sat at my desk and prayed very specifically, asking God to make it abundantly clear whether he wanted me to continue writing the novel. I promised God that if he asked me to lay it aside I would do just that.

Only about an hour later I received a curious email. It was from a teacher at a local Christian High School where I had previously taught Biblical Studies and Studies of Religion for 15 years. She informed me that she had set an assignment for her gifted students asking them to research the scientific challenges facing mankind in seeking to travel the vast distances to other solar systems and eventually colonise other planets. Her students needed to describe each of the challenges and evaluate the various proposed future technologies that are currently being researched and developed in order to overcome those challenges. At the end of her lengthy explanation of the student's assignment task, the teacher then asked me if I would be interested in coming and speaking to her class about these issues or, at the very least, did I have any helpful information that I could send to her students?

I was absolutely staggered by her email, for several reasons.

Firstly, I had always been a Bible teacher while at that school. I had never taught science or anything remotely connected with cosmology or space travel (although I had given a brief staff devotional talk many years previously about the wonders of the universe).

Secondly, the challenges of space flight and the colonisation of exoplanets was EXACTLY the topic of my science fiction novel. In fact, I had spent many, many hours researching exactly those issues in order to start writing the novel.

Thirdly, the teacher who sent me the email could not have had ANY idea that I was writing such a novel, because I had told NO ONE at that stage.

Fourthly and finally, the timing of her email was incredible: it arrived just an hour after praying for a clear sign from God. In fact, I can tell you that in my 60 years of life, I have NEVER received an invitation to speak about interstellar space travel EXCEPT on the very afternoon when I asked God to confirm if he wanted me to continue writing my novel about interstellar space travel! Once again, the statistical probability of this 'coincidence' must be staggeringly small.

As I reflected on the extraordinary timing of the email, I

concluded that God was telling me that he was happy for me to continue writing the novel – which I finished and published soon after, along with, eventually, a further three novels in what I now refer to as my science fiction 'quadrilogy'. (The first novel is called, "The Stars That Beckon").

At the risk of boring you to tears, allow me to give you one further example of God's intervention in my life.

One weekend in 2001, my wife and I and our two kids were staying for a few days with my parents at Forster, Australia. Late one afternoon, my young daughter suddenly realised that her watch was missing. She remembered taking it off while we were at the beach earlier that day. She had placed it in her towel while she went for a swim but couldn't remember seeing it after that. We concluded that it must have fallen out onto the sand when she unwrapped the towel to dry herself. The watch had sentimental value and my daughter was very upset.

We immediately rushed back to the beach, which was only 5 minutes away. When we got there the sun was setting and the beach was completely deserted. The lifeguards had long since packed up and left, and there was literally no one on the beach. We stood looking around trying to gauge roughly where our towels had been, but it was impossible to tell without the lifeguard flags with which to orient ourselves. We ended up standing in the middle of a vast stretch of beach staring hopelessly at the sand around us. I remember saying to my wife, *"It's hopeless. We'll never find it."* My wife, however, refused to give in. *"Let's pray and ask God to help us,"* she suggested.

I remember thinking, *"I'm sure God has much more important issues to be dealing with than finding a teenage girl's watch!",* but I didn't dare say it aloud. Not wanting to seem unspiritual, I agreed to lead our family in a brief prayer. We held hands and I simply asked God if he would help us find the watch. I must be totally honest with you at this point: I did not believe that God was going to answer that prayer. Even as I prayed, I was already starting to formulate in my mind a logical, reasonable explanation to give to my wife and children as we drove home empty-handed. I would explain to them that God doesn't always answer our prayers the way we would like. I would explain that God is much more concerned with transforming us and changing us on the inside than in helping us keep track of our possessions.

I was completely unprepared for what happened next.

As soon as I finished praying and we all said, "*Amen*", my wife knelt down in the sand, plunged her hand into a drift of sand and squealed. I thought she had been bitten by a crab! She hadn't. She drew her hand out of the sand and held up my daughter's watch! It had been buried under about 10cm (4 inches) of sand. It was literally the very first plunge of my wife's hand. In all that vast, empty beach, we had somehow come to stand precisely where our towels had been and, even more amazingly, my wife's hand had zeroed in on the watch like a radar-guided missile. To say that I was surprised would be a complete understatement.

Someone could claim that finding my daughter's watch in that way on the very first plunge of a hand was simply a fluke – a product of brute chance. But what kind of fantastically small probability are we talking about? We had absolutely no idea where our towels had been. For all we knew we could have been 50 or even 100 metres away from our previous spot on the beach.

I could multiply these stories a hundredfold – instances from my own life and the life of my family where we have seen God's miraculous hand at work. Only very occasionally have they been spectacular miracles, like the healing of my leg. Mostly, in fact nearly always, they have been much more mundane things involving phone calls, emails, watches and even tinier details of everyday life that God mysteriously "tweaks" when we ask for guidance or help of some kind.

Of course, the problem with testimonies of miracles such as these, is that you have no way of corroborating them. In the case of the healing of my leg at the beach, you weren't there. You didn't see my original injury. You didn't feel the pain. You didn't witness the healing. All you have is my word for it. For all you know, I could be making up that whole story. I could be making up *all* of these stories. And I can't help you with that. I don't have video footage that I can post on Facebook. I don't go through life with a video camera strapped to my head just in case a miracle happens. All I can do is tell my stories and assure you that I am not a nutter. I am not a person who is given to gross exaggeration or is carried away by emotions and hype. In fact, those who know me well will testify that I am a very logically-driven person who analyses everything carefully and demands unequivocal evidence before believing something. I am not easily fooled or self-deluded.

Over the years I have come across hundreds of testimonies of miracles, big and small, from sensible, logical people just like me. And while I concede that there will always be spurious claims of miracles by gullible, self-deluded people (and also by deliberately deceptive people!), it is the testimonies of miracles from sensible, reliable people that carry a great deal of weight. I have spoken with people who have had verified cancer, huge cancerous masses that were clearly evident on MRI scans, which disappeared after prayer for healing, with subsequent MRI scans just a week later completely clear. I have spoken with people whose doctors could not explain their healing but could verify its reality.

And there are not just a few of these people. There are millions of such people, all over the world, who testify that they have had some kind of tangible experience of God's miracle-working power.

I don't intend, in this book, to elaborate a detailed theology of healing or miracles. That is another discussion for another time (and one that I have dealt with in a previous book, *Making Sense of the Bible*). I am simply pointing out that the hand of God is sometimes evident in our world as he intervenes in people's lives in miraculous ways.

NATIONAL MIRACLES

Not all testimonies of miracles are personal. Sometimes God shows his hand at a national and international level.

The evacuation of the Allied army from Dunkirk in May 1940, is the story of a miraculous answer to prayer. On May 10th, 1940, Hitler unleashed a military onslaught on France and Belgium. Within days the Allied forces found themselves with their backs to the sea and hemmed in by enemies. They were vastly outnumbered and under-equipped and were facing complete annihilation. Winston Churchill prepared a press statement to announce to the public an unprecedented military catastrophe involving the capture or death of over a third of a million soldiers - almost the entire British army.

But it didn't happen. On May 23rd, King George VI requested that the following Sunday be observed as a National Day of Prayer. On that Sunday, the entire nation devoted itself to prayer in an unprecedented way. Photographs of that day show overflowing congregations in places of worship across the land. Long queues formed outside cathedrals. Every church throughout the land was filled to capacity and more, with crowds overflowing onto the footpaths and surrounding grounds. England had never seen a day like it. Almost the entire nation cried out to God in prayer, asking for his miraculous intervention for their 'boys' trapped in France and Belgium. The same day an urgent request went out for boats of all sizes and shapes to cross the English Channel to rescue the besieged army, a call ultimately answered by around 800 vessels.

As a result of that day of prayer a number of miraculous events unfolded:

Firstly, in a decision that infuriated the German generals and still baffles historians, Hitler ordered his army to come to a halt. Had they pressed on and continued to fight, the complete destruction of the British and Allied forces would have been inevitable, and the war would have taken a different, darker and more terrible path. Yet for three days the German tanks and soldiers stood idle while the evacuation unfolded.

Secondly, God miraculously 'tweaked' the weather. Terrible weather on the Monday and Tuesday grounded the German Luftwaffe, allowing Allied soldiers to march unhindered to the beaches without being strafed by the German fighter planes.

Thirdly, on the following day, Wednesday, the day of the evacuation, the sea was extraordinarily calm, despite the storm that had grounded the Luftwaffe over the previous days. This allowed all the tiny private English yachts and other small vessels to sail right into shore to load the evacuees.

By the time the German Army was finally ordered to renew its attack, over 338,000 troops had been snatched from the beaches; a third of a million! Many of them were to return four years later to liberate Europe.

Some people might say it was all a coincidence, but I think not.

It certainly wasn't considered a coincidence at the time. Sunday, June 9th, was declared a National Day of Thanksgiving, with the nation going to church again to thank God for his deliverance, and Winston Churchill himself referring to the evacuation as the 'The Miracle of Dunkirk'.

If you are looking for evidence of God's existence, there is plenty to be found in everyday life. The testimony of people's encounters with God and their experience of his miraculous intervention at both personal and national levels provides compelling evidence of a God who is both powerful and personal. Stories such as the ones I have recounted in this chapter should not surprise us. After all, if God exists, it is entirely reasonable to expect to encounter his influence at some level and at various times in our world.

For my own part, my experience of God's intervention in my everyday life is a constant confirmation of not only his existence, but also of his loving, intimate, personal presence. This is why, when I am asked if I believe in God, my response is often something like, "*I don't just **believe** in God; I **know** there is a God, because I experience him in my life.*" I could no sooner deny the existence of God than deny the existence my wife and family. You see, for me, the question of God's existence is no longer an abstract or theoretical one, because I have encountered him personally and powerfully.

The Bible says:

> "*Taste and see that the Lord is good.*" (Psalm 34:8).

This is an invitation; an invitation to move beyond mere intellectual enquiry and to open yourself to encounter God personally. You see, intellectual enquiry will only take you so far. You will reach a point where, having examined all the evidence, you might conclude that there is a possibility that God exists. You may even conclude that it is highly **probable** that God exists. You may reach a point where you might say, "*Yes, I now believe in God.*" Excellent! But at that point, God's existence will still be an intellectual exercise for you. He will still be a distant Deity whose existence you concede but whose presence you have not yet encountered at a personal level.

That is when you must take the next step.

And that is the topic of the next chapter.

10
PASCAL'S WAGER

In this book, I have presented you with seven areas of evidence that point to the existence of God:

1. Evidence of a supernatural cause for the origin of the universe
2. Evidence of a supernatural cause for the origin of biological life and consciousness
3. Evidence of intelligent design throughout the universe
4. The existence of objective moral values which are only possible if God exists
5. Evidence of God's intervention throughout history
6. Evidence supporting the deity and resurrection of Jesus Christ
7. Evidence of personal experience and miracles

If we consider this entire portfolio of evidence, it seems to me that it would take much more faith to DISBELIEVE in God than to believe in him. As each piece of evidence is added to the argument, the case for God's existence becomes increasingly compelling until, in my opinion, it becomes utterly convincing.

The simplest and most logically consistent explanation of *all* the evidence is that a Creator God has brought our universe into existence and continues to interact with it.

The person who attempts to explain away the evidence, however, requires a great deal of faith indeed. The atheist who rejects God as the most likely explanation is left with an unsolvable puzzle. If not God, then what? How else can all the evidence be explained? Atheists either have to resort to increasingly speculative and even bizarre alternative explanations for the various pieces of evidence, or else they must concede that they simply don't have a viable explanation. This is especially true in regard to the first three areas of evidence. The atheist is left saying, *"I have no explanation for the origin of the universe, the origin of life and the indisputable signs of intelligent design, but I don't accept the God-explanation."* In effect, the atheist is left believing in an alternate explanation that he cannot yet perceive. He is effectively saying, *"Something else caused all this, but I have no idea what it was."* This is why I maintain that it takes MORE faith to disbelieve in God than to believe in him.

What about you? How do you respond to the evidence? As we saw in Chapter 1, former atheist, Dr Antony Flew, stated, *"We must follow the evidence, wherever it leads."* For him, it led to a complete renunciation of his previous atheism. But not everyone is as brave in following the evidence to its logical conclusion. Some people look at the evidence and see where it is heading but draw back from the brink. They don't like where it is leading them. They don't want to believe in God because to do so would ruin their independence. They don't want to be accountable to a Creator, so they reject the most logical explanation in favour of anything, no matter how fanciful or nebulous, that will allow them to maintain their sovereign independence.

As Winston Churchill once stated,

> "Men occasionally stumble over the truth, but most of them pick themselves up and hurry off as if nothing had happened."[1]

So, I repeat: how do YOU respond to this evidence? Are you bold enough to follow it wherever it leads?

Let me reiterate something that I pointed out in Chapter 1: we are talking about 'evidence' not 'proof'. If you are waiting until you are completely and utterly convinced, until God's existence is proved in an absolute, undeniable sense, you will never reach a decision. Because, as we have already discussed, absolute proof for just about everything is simply not possible. In most things in life, we have to act in good faith based upon the weight of evidence that establishes something beyond reasonable doubt. And, in my opinion, the existence of God has been demonstrated beyond reasonable doubt.

PASCAL'S WAGER

Blaise Pascal was a 17th century French philosopher, theologian, mathematician and physicist who proposed an interesting argument for having faith in God. Recognising the impossibility of ever attaining absolute proof of God's existence, he argued that believing in God was a much safer bet than not believing in him. His argument was fairly complex but, in essence, it can be expressed as follows:

- If you believe in God and follow his commands but it turns out that God *doesn't* exist, you have lost very little (perhaps some pleasures or 'sins' you might have otherwise enjoyed).
- But if you disbelieve in God and, consequently, disobey

his supposed commands, and it turns out that he *does* exist, your loss is *infinite* (an eternity separated from him and from heaven).

Pascal concludes, therefore, that it is a much safer wager to be a Christian. In fact, he reasons that considering the huge potential losses of unbelief and the very negligible potential losses of belief, being a Christian is the most sensible and logical decision. He also points out that no one can escape the necessity of making this decision. You can't opt out of the game. If you don't decide to be a Christian, then you have effectively decided against God. In this sense everyone is making a 'wager'.

It's an interesting way of looking at it, isn't it? But I disagree with Pascal on the issue of the minimal potential loss attached to belief. I argue that there is *no* potential loss at all – that living a Christian life is BETTER by far than not doing so – that there is no potential loss whatsoever – even if it turns out that God doesn't exist.

Consider, for example, the issue of sex outside of marriage. The Bible teaches that God has ordained that sex is for marriage only and that those who follow him should not engage in sex outside of marriage. Consequently, my wife and I have been completely faithful to each other, and neither of us has ever had any other sexual partners. A hedonist (someone who lives for the pursuit of pleasure) might argue that we are missing out on the excitement of having an affair or two, or that we missed out on fun by not having sex before marriage. They would argue that our Christian lifestyle is hindering our experience of pleasure. I disagree. My observation of the result of pre-marital sex and extra-marital affairs is that the eventual pain and heartache that they cause far outweighs any transient pleasure they might bring. Our Christian commitment to a faithful marriage actually fosters security, trust, love and intimacy – all of which would be damaged or destroyed if either one of us had an affair. In the case of people who have led sexually active lives prior to marriage, I have often noted ongoing negative consequences in their later lives, including trust issues, ongoing infidelity, guilt, depression, self-esteem issues, sexual dysfunction, as well as persistent health issues from incurable sexually transmitted infections. After a lifetime of counselling people and seeing the devastation caused by pre-marital and extra-marital sex, I have absolutely no doubt that following God's commands in this area of life results in the BEST possible life. I have not lost anything by obeying God.

The same can be said of any so-called 'pleasure' that I might be forsaking in order to follow Christ. In fact, now that I think of it, I can't really identify *any* good thing that I am giving up in that sense. Drunkenness and drug use don't hold any appeal for me, because they, like all other sins that the Bible warns us against, actually lead to LESS pleasure in the long run. They are DESTROYERS of joy and contentment and health. Similarly, in following Christ, I turn aside from unforgiveness, choosing instead to forgive others. This might seem like I am losing something, giving up the 'right' to seek revenge, but it turns out that carrying grudges and seeking revenge would actually hurt me in the long run.

You see, it turns out that the Christian lifestyle results in a much BETTER life, because the things we are called to give up are things that would actually hurt us and make us miserable.

So I would alter Pascal's Wager to read as follows:

- If you believe in God and follow his commands but it turns out that God DOESN'T exist, you have lost NOTHING – in fact, you've had a *better* life.
- But if you disbelieve in God and it turns out that he

DOES exist, your loss is INFINITE (an eternity separated from your Creator and from heaven).

This makes the decision to believe in God and follow him even more logical. In fact, it makes the decision to *not* follow God completely *illogical* doesn't it? If following God gives you a better life and potentially brings an infinite reward, but rejecting God results in a life that is really no better (and arguably worse) and potentially results in an *infinite* loss, why would anyone choose the latter?

To be clear, Pascal's Wager isn't an argument for God's existence, but it is a very strong argument for placing your faith in God once you have *examined* the evidence. Pascal is effectively asking the question, *"What's stopping you from placing your faith in God? What have you got to lose?"*

The answer is NOTHING.

There is actually NOTHING to lose and EVERYTHING to gain.

But what does having faith in God look like? What is actually involved?

TAKING THE NEXT STEP

The focus of this book has been to help people come to a *belief* in God; hence the title, *7 Reasons to <u>Believe</u>*. Perhaps you have been helped in this way. Perhaps you have reached the conclusion that God exists, and you can now say *"I believe in God."* That's a momentous first step, but it's important that you don't stop there. The Bible tells us that God is not merely looking for '*belief*', he is looking for '*faith*'. There's a big difference. Belief is passive, but faith is active. Belief is mere intellectual assent, whereas faith involves an active commitment to something or someone.

An illustration might help.

I believe that having a cold shower every day is beneficial for my health (I really do!). Research suggests that immersing your body briefly in cold water energises your metabolism, stimulates your immune system and does a whole lot of other helpful stuff (am I being too technical?). I really do believe in the practice of having a cold shower every day.

But I don't do it. That's right, you heard me: I don't have cold showers. They're too cold! You see, I believe in the value of cold showers, but I'm just not willing to commit myself to doing it. I have a belief in something, but because that belief hasn't translated into action – into commitment – my belief does me no good. I derive absolutely no benefit from that belief; it is mere intellectual assent.

A lot of people are like that in regard to God. They believe in God. They may even believe in Jesus. They may believe in the Christmas story of God becoming flesh. They may believe in the Easter story of Jesus dying on the cross for our sins and rising again to bring us new life. But that belief is mere intellectual assent. It does them no good, because they are not willing to commit themselves to that belief and to the God who is the object of that belief. Such a person believes in God and Jesus in the same way I believe in cold showers: neither of us is actually willing to change our lives for that belief.

Significantly, Jesus never said to anyone, *"Believe in me."* Instead, the Gospel writers record 22 instances of Jesus saying, *"Follow me."* A Christian is a *follower* of Jesus, rather than a mere believer.

Unfortunately, the concept of following someone has been

undermined by social media in recent times. It's relatively easy to follow someone on Facebook or Instagram. You simply tick a box and you are now one of that person's followers. But you don't have to agree with everything they say, and you certainly don't have to do everything they demand. You can follow someone on social media without it changing your life in any way.

Not so with Jesus. The kind of 'followership' that Jesus is talking about is the kind that demands total commitment. It is a commitment to live the way he commands you to; abiding by his standards and implementing his teaching. In following Jesus you are seeking to be like him and to live as he would live. You can't follow Jesus like you follow someone on social media; he doesn't give us that option.

In Mark's Gospel, chapter 8 verse 34, we read:

> "And calling the crowd to him with his disciples, he said to them, 'If anyone would come after me, let him deny himself and take up his cross and follow me.'"

Here we see the full extent of the kind of 'followership' that Jesus requires. It is a followership of self-denial and sacrifice. This doesn't mean adherence to some sort of stringent, ascetic lifestyle of physical deprivation. The denial and 'death' ("take up your cross") that Jesus is referring to is the death of your self-rule, the demise of your autonomy. Following Jesus means making him the Captain of your life. He gets to call the shots now: he gets to determine your values and guide your actions. Following Jesus means handing the steering wheel of your life over to him. This is what the Bible means by calling Jesus, 'Lord'. It was the first century term for 'master'.

Of course, this doesn't imply perfection. The followers of Jesus don't get it right all the time. We don't follow Jesus perfectly. We occasionally stumble and fall. We don't always perfectly reflect his standards. We don't always perfectly obey his commands. We still sin from time to time. But there is forgiveness when that happens. That is why Jesus had to die on the cross – to pay for our sins so that we could be forgiven; so that he could lift us up when we stumble, put us on our feet again and say, *"Let's try that again, shall we?"* Following Jesus will never mean obeying him **perfectly**, but it does mean being fully committed to **trying** and to the process of **growing** in obedience.

All this talk of commitment and obedience and self-denial might lead some to believe that following Jesus is a life of austerity and bleak conformity to a harsh set of rules. But this is not so at all. Those who follow Jesus soon find that they experience a degree of joy that was never there before. Jesus promised, *"I have come that you may have life, and have it to the full" (John 10:10)*. This is not an empty promise. Anyone who has become a follower of Jesus will testify that their life is fuller and richer as a result, and there is a very good reason for this. We were created to know God, and when we enter into relationship with him, through his Son Jesus, something that was missing is finally restored. As Augustine of Hippo wrote, in the 4th century:

> *"You have made us for yourself, O Lord, and our hearts are restless until they find their rest in you."* [2]

There is a moving incident in the life of Jesus that demonstrates the point that Augustine is making. In Chapter 4 of John's Gospel, we read of an encounter between Jesus and a Samaritan woman who has come to draw water from a well on the outskirts of a small town called Sychar, in Samaria. There is a fascinating backstory to this incident, regarding the racial tension that existed between Jews and Samaritans, that makes Jesus' interaction with the woman surprising and counter-cultural. But what is much more relevant to our present discussion is the woman's dysfunctional life and her thirst for something more meaningful, which becomes clear as her interaction with Jesus unfolds.

Jesus begins the encounter by asking the woman if she would share with him some of the water that she has drawn from the well, as he is thirsty after a long walk. Very quickly, however, Jesus moves their conversation into 'deeper waters'. He tells the woman that he can give her "living water" and says to her:

> "Whoever drinks the water I give them will never thirst. Indeed, the water I give them will become in them a spring of water welling up to eternal life." (John 4:14)

The woman is initially confused by this statement, thinking that Jesus is referring to some kind of physical water. But, of course, Jesus is speaking of something much more profound. He is offering to fill up the emptiness inside her and satisfy her thirst for meaning and fulfilment. Sensing the woman's confusion, Jesus decides to highlight the woman's unhappiness and restlessness. The following intriguing exchange takes place:

> He told her, "Go, call your husband and come back."
>
> "I have no husband," she replied.
>
> Jesus said to her, "You are right when you say you have no husband. The fact is, you have had five husbands, and the man you now have is not your husband. What you have just said is quite true." (John 4:16-18)

In this casual but remarkable display of divine insight, Jesus strips away the façade that the woman is presenting to him and lays bare her desperate search for happiness and meaning – a search that has led her through a succession of failed relationships.

The encounter ends with the woman running back to her town, excitedly proclaiming that she believes she has found the Messiah (a Jewish term for Saviour).

She is right. Jesus is, indeed, the Saviour. He offers to wipe away our sin, forgive our past and give us a fresh start in life – a life connected to the God we were created to know. And for those who accept Jesus' offer, they discover a level of profound joy and fulfilment that they didn't know they were missing.

This is what Jesus meant when he said:

> "I have come that you may have life – life in all its fullness" (John 10:10).

And this is precisely what Augustine meant when he stated:

> "You have made us for yourself, O Lord, and our hearts are restless until they find their rest in you."[3]

AN INVITATION TO ENCOUNTER GOD

You see, ultimately, God calls us to not merely believe in him, but to open our hearts to encounter him. And, as I have already explained, this necessarily involves submitting our lives to him and becoming a true follower.

In my first book, *"Finding God When He Seems To Be Hiding"*, I described an encounter I once had with a man who had previously belonged to a criminal motorcycle gang. It is worth repeating here, because it beautifully illustrates what God requires from all of us.

One Sunday night in 2017, I had a speaking engagement at a local church outreach dinner where a large number of non-church people were in attendance. I began by dealing with some of the common objections to Christianity and then presented a basic explanation of the Christian message, including Jesus' wonderful offer of forgiveness and a new start in life.

After my talk I was approached by a Christian man in his late 50's, called Ron, who shared the story of how he had sought and found God when he was in his 20's. He had been a member of an outlaw biker gang and had been living a wild life. He recounted stories of some of the terrible things he had done, and how dark and violent his heart had been. On one of his many drunken bike rides he had a horrendous crash and was in a coma for three days, yet even such a serious incident did not cause him to stop and evaluate the path he was on. Eventually his life of crime and violence caught up with him. He was arrested, charged and convicted for armed robbery and attempted murder. He was sentenced to seven years in prison and was sent to the maximum-security section of one of Australia's largest prisons.

At this point in the story he simply said to me, "That's where I found the Lord". I asked him, "How did that happen?". In my mind's eye, I pictured him being befriended by the prison chaplain, starting to attend church services, asking lots of questions, beginning to read the Bible and gradually developing a faith that ultimately blossomed into a genuine commitment to Christ as Lord and Saviour. I was wrong! I'll let Ron tell you how it really happened in his own words:

"It was my second night in prison. I was alone in my cell and I was absolutely terrified. I was facing seven years in a maximum-security prison and I knew I couldn't do it. I remember thinking, 'I'm not gonna make it.' I was in a really bad way. I realised what a mess I'd made of my life and I felt like I was at the bottom of a deep black hole with no way out. I fell to my knees on the concrete floor and I raised my arms to heaven and I cried out to God with all my heart. My prayer went something like this; 'God I don't know whether you're there or not, but if you are, please help me! I can't do this on my own! Please help me! Please come into my life! Please rescue me! I need you! I'm sorry for what I've done. Please help me!' Immediately God turned up in a big way. I physically felt Him wrap His arms around me and I was completely overwhelmed by a feeling of warmth and love and joy – like nothing I had ever experienced before in my life. The feeling of God's presence coursed through my body from head to foot. I started thanking and praising Him, and I was so overwhelmed by His presence that I was sobbing uncontrollably from sheer joy."

Wow! What a testimony! Ron has been a Christian now for over 30 years. After prison he married a wonderful Christian lady and worked as a youth minister, then as a counsellor in a drug rehabilitation facility. He is now retired and if you were to meet him you would have a strong impression of someone who has walked closely with the Lord for many years. In fact, as I spoke with him, I couldn't help wondering why the church hadn't asked him to do the talk that night instead of me!

Ron's story illustrates an important truth. God invites us to encounter him, so that we might be transformed. It's a stunning fact: the God of creation, the One who made this whole amazing universe, invites us to enter into a life-changing relationship with him. We are invited to not merely **believe** in God, but to **know** him and be transformed by him.

. . .

THE NEXT STEP IS UP TO YOU

So, you see, believing in God and Jesus is not the end of the journey. It merely brings you to the point where you are ready to consider God's claim on your life. He made you and he now calls you into relationship with himself. He calls you to be a *"follower"*, not a mere *"believer"*.

The Bible says:

> *"To all who **received him** [Jesus], to those who believed in his name, he gave the right to become children of God."* (John 1:12)

More than merely *believing* in Jesus, we are called to *"receive him"*, to submit our hearts and lives to him as our new master, in order that we might experience the fullness of life that we were created to find.

And, as Pascal's Wager argues, what have you got to lose?

I sincerely hope and pray that this book has helped to strengthen your belief in God. More importantly, I hope it has motivated you to do something about it. The God who created you and loves you enough to send his Son to die on a cross for you, now calls you back into relationship with himself. The next step is up to you.

Go on.

Take that step.

You've really got nothing to lose.

A HELPFUL PRAYER

"God, I acknowledge you now as my God and my Creator. I ask you to forgive me for living without any meaningful reference to you. I ask your forgiveness for the many times I have defied your commands. Thank you, Jesus, for dying on the cross and rising again to pay for my sins and purchase my forgiveness. I open my heart to receive you as my Master. Please come into my life, fill me with your presence, and strengthen me to follow you all the days of my life. Amen."

OTHER TITLES BY KEVIN SIMINGTON

FINDING GOD WHEN HE SEEMS TO BE HIDING

"*Finding God When He Seems to Be Hiding*" provides intelligent answers to the common questions and objections that are often roadblocks in people's journey towards faith. If God exists, why is there so much suffering in the world? What about all the killing in the Bible? How can a loving God send people to hell? Is the Bible reliable? What evidence is there for the resurrection of Jesus? What about evolution? Hasn't science and evolution disproved the existence of God? How can God permit abuse and religious violence?

This book addresses these and other common questions with remarkable clarity and provides answers that move beyond the standard, glib responses that are often proposed.

"*Finding God When He Seems To Be Hiding*" is available in print or as an eBook from SmartFaith.net, Amazon and all major online retailers.

MAKING SENSE OF THE BIBLE

This book will change the way you read the Bible!

"*Making Sense of the Bible*" is a comprehensive guide to understanding and interpreting the Bible. It explores the remarkable journey of the Bible, from original text to modern translation, and will assist you to develop a more mature, complex understanding of the nature of its divine inspiration. It examines the many complex cultural and contextual issues that are essential in order to accurately apply the Bible's message. These include the difference between the two covenants, the nature of progressive revelation, the pre-Christian context of the Old Testament, and the necessity to read the whole Bible "Christologically" - through the lens of Christ's person and work.

What sets "*Making Sense Of The Bible*" apart from similar books is its intensely practical nature. Commonly misinterpreted doctrines are explored in detail, and important principles of interpretation are applied. A large range of key biblical doctrines are examined in detail.

This book is a must for ordinary Bible readers and serious students alike!

"*Making sense of the Bible*" is available in print or as an eBook from SmartFaith.net, Amazon and all major online retailers.

NO MORE MONKEY BUSINESS:
EVOLUTION IN CRISIS

"*No More Monkey Business*" is a concise, easy-to-read summary of the overwhelming and rapidly accumulating scientific evidence against evolution. Written with wit, and using simple layman's language, yet brimming with incontestable scientific evidence, this book highlights the huge problems now facing

Darwin's original theory. Each chapter is full of fascinating scientific facts and discoveries which now directly contradict Darwin's naïvely simplistic theory proposed more than a century ago. *"No More Monkey Business"* documents the abandonment of the theory of evolution by a growing tide of the world's leading scientists, as well as the startling declaration by several recent scientific conferences that the theory of evolution can no longer be considered to be scientifically tenable. This book will challenge those who have unthinkingly assumed evolution to be a proven fact and will enable Christians to defend their faith with confidence.

"No More Monkey Business" is available in print or as an eBook from SmartFaith.net, Amazon and all major online retailers.

RETHINKING THE GOSPEL

"Rethinking the Gospel" is a profoundly challenging exploration of the modern church's proclamation of the gospel. It examines an element of the gospel that has been largely ignored or under-emphasised since the start of the Reformation in the 1500's. In many churches today, this aspect of the gospel is almost entirely missing from public preaching, and its absence has had a significant impact upon the kind of disciples that the modern church is producing. Martin Luther's important rediscovery of the biblical doctrine of salvation by grace through faith alone and his complete denunciation of good works as a means of self-justification has shaped evangelical preaching to the current day. But in the modern church's fervour to present salvation as a free gift, it has under-emphasised the necessity of repentance and ongoing submission to the Lordship of Christ – a theme that was a central tenet of Jesus' teaching. The kind of faith that Jesus demanded was not the passive, easy-believism of today's preaching but, rather, a robust faith of self-denial, sacrifice and daily obedience to Christ as Lord. It was a faith that resulted in a deeply transformed life.

"Rethinking the Gospel" will challenge you to re-examine your understanding of the gospel in the light of Jesus' consistently confronting teaching on the relationship between faith and obedience.

One reviewer commented: *"Every church pastor, preacher and Christian should read this book! It has transformed my understanding of the gospel."*

Another reviewer commented: *"This is a devastating and eye-opening commentary on the blight that has infiltrated the modern church. It is a wake-up call that desperately needs to be heard."*

"*Rethinking the Gospel*" is available in print or as an eBook from SmartFaith.net, Amazon and all major online retailers.

THE LITTLE BOOK OF CHURCH LEADERSHIP

"*The Little Book of Church Leadership*" is a little book with BIG ideas.

There is a crisis of leadership within the modern church. Not an *absence* of leadership. The modern church has plenty of leadership; just not the right sort. The kind of leadership that has evolved in many churches today is a long way from the leadership that was taught in the New Testament and practised by the first century church. Many churches have absorbed a philosophy and style of leadership that is more attuned to the business world than to the Bible. This book is a call to seriously re-evaluate the church leadership style that has developed in recent years and return to the patterns and principles of church leadership as outlined in the New Testament.

"*The Little Book of Church Leadership*" is available in print or as an eBook from SmartFaith.net, Amazon and all major online retailers.

WELCOME TO THE UNIVERSE

How big is our solar system? Our galaxy? The universe? Does extra-terrestrial life exist? How unique is Earth? Will we ever be able to travel to other stars? How realistic are the science fiction accounts of space travel?

"*Welcome To The Universe*" addresses these and many other issues of cosmic proportion. With stunning photographs and mind-boggling facts, "*Welcome To The Universe*" provides a fascinating glimpse into the wonders of the universe and the many challenges of space travel. It is the perfect 'pocket sized' compendium for budding astronomers and armchair lovers of science and science fiction.

"*Welcome To The Universe*" is available in print or as an eBook from SmartFaith.net, Amazon and all major online retailers.

SOMEONE ELSE'S LIFE

"A page-turning thriller by a master story-teller!"

What if you weren't who you thought you were? ... And people will kill to stop you finding out!

Much more than a simple detective story, this is a complex portrayal of a good man who is pushed to extraordinary limits.

A mysterious case of identity switching turns deadly when struggling private investigator, John Targett, becomes involved. As John seeks to unravel one mystery, he is also forced to deal with an escalating menace when he becomes the target of a vicious gang whose path he has crossed. As the twin plots intertwine and the threats escalate, John is forced to take extreme measures to protect his daughter and fight for his own life. Plagued by his own demons and trying to raise his daughter alone, this is a beautifully crafted story of the lengths to which one man will go to protect those he loves. At times tender, filled with sparkling wit and peppered with edge-of-your-seat action, this is a multi-facetted mystery that will satisfy on many levels.

REVIEWS:

"An incredible thriller with the perfect twist! I adored this book. John Targett is my newest character crush! **Someone Else's Life** delivers on every front. It's delightful, witty, dangerous, and thought-provoking. The danger level is high throughout the novel, constantly raising the stakes and potentially making the reader breathless as events unfold. It's thrilling, and absolutely ends on the best possible note." Kat Cohen, Reviewer.

"Wow - this book is so much FUN! Great characters and a high-octane story that rips along until the twist you won't see coming on the last page. John Targett is an impressive new hero - tough and yet tender; highly skilled and often very funny. His laugh-out-loud interactions with his teenage daughter were one of my highlights." Darren Box, Amazon Review.

Available from: kevinsimington.com and Amazon

THE STARPATH SERIES

A science fiction adventure series that is consistently receiving 5-star reviews around the world!

A dying world.
A desperate mission.
An unlikely hero.

"Incredibly well written, intelligent science fiction, by an author who really knows how to tell a story."

"I was hooked from the first page. The story is gripping and moves at a cracking pace. I also loved that there was humour and romance as well as edge-of-your-seat drama."

Book 1: THE STARS THAT BECKON
Book 2: THE STARS THAT BEND TIME

Book 3: A PATH THROUGH THE STARS
Book 4: Eden Rising

Available from: kevinsimington.com and Amazon

CONNECT WITH KEVIN SIMINGTON

Non-Fiction books and resources: SmartFaith.net

Fiction books: kevinsimington.com

Facebook: https://www.facebook.com/ReflectionsKev/

ABOUT THE AUTHOR

Kevin Simington is a theologian and apologist who is passionate about helping Christians grow deeper in their faith. He spent 31 years in Christian ministry, as a church pastor and a Christian educator. He is now a full time author and speaker. His website, SmartFaith.net, and Facebook page, "Reflections on Faith and Life", provide valuable resources for defending the Christian faith and equipping Christians. Kevin's weekly blog, available through his website and Facebook page, provides incisive commentary on social issues, theology, apologetics and ethics, and is read by thousands of people worldwide. He also writes for "My Christian Daily", an international Christian magazine.

NOTES

1. The Nature And Limitations Of Evidence

1. Isham, C. J. (1997). Creation of the universe as a quantum process. In Russell, R. J., Stoeger, W. R., and Coyne, G. V., editors, Physics, Philosophy, and Theology: A Common Quest for Understanding. Vatican Observatory, Vatican City State/University of Notre Dame, third edition.
2. Antony Flew, "There Is A God", Harper Collins, 2007, p. 127
3. Antony Flew, "There Is A God", Harper Collins, 2007, p. 136
4. Antony Flew, "There Is A God", Harper Collins, 2007, p.136
5. Antony Flew, "There Is A God", Harper Collins, 2007, p.136

2. Removing The Elephant

1. P.S. Moorhead & M.M. Kaplan, "Mathematical Challenges to the Neo-Darwinian Interpretation of Evolution", The Wistar Institute Symposium Monograph No.5, Philadelphia, P.A., Wistar Institute Press, 1967).
2. Barbara J. Stahl, "Vertebrate History; Problems in Evolution", New York, McGraw-Hill, 1973.
3. Roger Lewin, "Science" Journal, Vol. 210(4472), 1980, pp.883-887
4. Michael Denton, "Evolution: A Theory in Crisis", Bethesda, Adler & Adler, 1986.
5. Lee M. Spetner, "Not By Chance: Shattering The Modern Theory of Evolution", New York, Judaica Press, 1997.
6. Werner Gitt, "In The Beginning Was Information", Green Forest, A.R., Master Books, 2006.
7. Cercle D'études Scientifique et Historique website, also quoted in "Evolution; Fact or Belief?", Creation Science Foundation.
8. Susan Mazur, The Altenberg 16: "An Expose of the Evolutionary Industry", Berkeley, CA, North Atlantic Books, 2010.
9. John F. Ashton, "In Six Days: Why Fifty Scientists Choose to Believe in Creation", Green forest, AR, Masterbooks, 2001.
10. John Ashton, "Evolution Impossible: 12 Reasons Why Evolution Cannot Explain the Origin of Life on Earth", Masterbooks, Green Forest, AR, 2012.
11. Casey Luskin's chapter, "The Top Ten Scientific Problems with Biological and Chemical Evolution" in the book, "More than Myth" (Chartwell Press, 2014).

12. E.H. Andrews, "God, Science and Evolution". Creation Life Publishing, 1981, p.90
13. E.H. Andrews, op. cit.
14. "The Pew Forum", conducted 2007, released 2008.

3. Evidence 1: The Origin Of The Universe

1. Debate between Bertrand Russell and Fr. F. C. Copleston, broadcast on BBC radio, 1948, transcript: http://www.biblicalcatholic.com/apologetics/p20.htm
2. https://history.aip.org/exhibits/cosmology/ideas/expanding.htm
3. https://history.aip.org/exhibits/cosmology/ideas/expanding.htm
4. https://en.wikipedia.org/wiki/Friedmann%E2%80%93Lema%C3%AEtre%E2%80%93Robertson%E2%80%93Walker_metric
5. https://en.wikipedia.org/wiki/Hubble%27s_law
6. https://en.wikipedia.org/wiki/Cosmic_microwave_background
7. http://www.hawking.org.uk/the-beginning-of-time.html
8. J. Trefil, *The Dark Side of the Universe*. Charles Scribner's Sons, Macmillan Publishing Company, New York, USA, pp. 3, 55, 1988.
9. Halton C. Arp, quoted in E.P. Fischer (Ed.), *Neue Horizonte 92/93—Ein Forum der Naturwissenschaften—Piper-Verlag*, München, Germany, pp. 112–173, 1993
10. https://creation.com/expanding-universe-2
11. http://blog.lege.net/cosmology/cosmologystatement_org.pdf
12. Lerner, E., Bucking the big bang, *New Scientist* 182(2448)20, 22 May 2004
13. "Big Bang Theory Busted by 33 Top Scientists", www.rense.com, 27 May 2004.
14. Al-Ghazali, quoted in Greg Dewar, "Advanced Philosophy and Ethics of Religion", Oxford University Press, Oxford. 2002, p.18.
15. Al-Ghazali, quoted in Greg Dewar, "Advanced Philosophy and Ethics of Religion", Oxford University Press, Oxford. 2002, p.18.
16. Debate between William Lane Craig and Quentin Smith, https://www.reasonablefaith.org/media/debates/does-god-exist-the-craig-smith-debate-2003/
17. https://tasc-creationscience.org/article/scientific-evidence-points-creator
18. https://en.wikipedia.org/wiki/Robert_Jastrow
19. James Clerk Maxwell; Perspectives on His Life and Work", Oxford University Press, 2014, p.274
20. Dr Hugh Ross, "The Creator and The Cosmos", Navpress, 2001, pp.108-112).
21. Dr Hugh Ross, "The Creator and The Cosmos", Navpress, 2001, pp.108-112).

22. Quoted by Hugh Ross, "The Creator and The Cosmos", Navpress, 2001, pp.108-112
23. Richard Dawkins interview, https://www.youtube.com/watch?v=of-8Q3HySjE&t=44m08s
24. Proposed by Paul Steinhardt and Alexander Vilenkin, in 1983. https://en.wikipedia.org/wiki/Eternal_inflation
25. https://en.wikipedia.org/wiki/Cyclic_model
26. Stephen Hawking, "Black Holes and Baby Universes", Random House Publishers, 1994.
27. Video Clip of interview between Ben Stein and Dr Richard Dawkins, "Richard Dawkins Believes Extraterrestrials Created Man." https://www.youtube.com/watch?v=AiVoS78lNqM
28. https://www.abc.net.au/news/science/2018-09-02/block-universe-theory-time-past-present-future-travel/10178386
29. https://www.theguardian.com/technology/2016/oct/11/simulated-world-elon-musk-the-matrix
30. https://www.theguardian.com/technology/2016/oct/11/simulated-world-elon-musk-the-matrix
31. Stephen Hawking, cited in *Michael Holden (2010-09-02)*. "God did not create the universe, says Hawking". Reuters. Retrieved 2010-10-17
32. Article on Americanscientist.org, March 2017, no longer available.
33. Isham, C. J. (1997). Creation of the universe as a quantum process. In Russell, R. J., Stoeger, W. R., and Coyne, G. V., editors, Physics, Philosophy, and Theology: A Common Quest for Understanding. Vatican Observatory, Vatican City State/University of Notre Dame, third edition.
34. John Leslie, Cosmological Investigations: Platonic Creation Theory, Oxford University Press, Oxford, 2003
35. Stephen Hawking, Black Holes and Baby Universes, 1993, New York, Bantam. P.172
36. "The Language of God", Francis Collins, 2007, Simon & Schuster, United States
37. Robert Boyle (2000). "The Works of Robert Boyle: Publications of 1674, p. 6
38. Scientific Autobiography and Other Papers as translated by F. Gaynor (1949), p. 184

4. Evidence 2: The Origin Of Life

1. Michael Behe, Darwin's Black Box: The Biochemical Challenge to Evolution, Simon & Schuster, 1996,
2. Tas Walker, "Irreducible Complexity and Cul-De Sacs", https://creation.com/irreducible-complexity-and-cul-de-sacs, 17 Dec 2016

3. https://www.youtube.com/watch?v=wJyUtbn0O5Y
4. Alex Williams, "Life's Irreducible Structure - Part 1: Autopoiesis", 2007, in Journal of Creation 21(2):109-115
5. Jonathan Sarfati, "Refuting Evolution 2", 2013, Creation Book Publishers, Powder Sprongs, GA, Ch. 10, point 14.
6. Michael Behe, "Darwin's Black Box: The Biochemical Challenge to Evolution", Free press, New York, 2006
7. Michael Behe, "Darwin's Black Box: The Biochemical Challenge to Evolution", Free press, New York, 2006, p.39
8. https://evolutionnews.org/2011/03/michael_behes_critics_make_dar/ See also: http://www.talkorigins.org/faqs/behe.html
9. http://www.ideacenter.org/contentmgr/showdetails.php/id/840
10. John F. Ashton. "Evolution Impossible", Green Forest, AR, Masterbooks, 2013, p.40
11. Opp cit, pp.40f
12. Clifford D. Sirnak, "Trilobites, Dinosaurs and Man", New York, St Martins Press, 1966, p.54
13. Isaac Asimov, "The Genetic Code", New York, The New American Library, 1962, p.92
14. Isaac Asimov, "The Genetic Code", New York, The New American Library, 1962, p.92
15. Thomas Heinze, "Did God Create Life? Ask a Protein.", https://creation.com/did-god-create-life-ask-a-protein
16. Hoyle, Fred (1983). The Intelligent Universe. p. 17. The Boeing 747 metaphor is reported in Nature, 294 (1981), p.10
17. Kenyon DH, Steinman G. Biochemical Predestination. McGraw Hill Text (January, 1969) ISBN 0-07-034126-5.
18. https://en.wikipedia.org/wiki/Dean_H._Kenyon
19. https://en.wikipedia.org/wiki/Dean_H._Kenyon
20. Kenyon, D., Foreword to What is Creation Science? by Henry Morris, San Diego, Creation-Life Publishers, pp. i-iii, 1982
21. "Unlocking the Mystery of Life", Illustra Media, 2010
22. Charles Darwin, "The Origin of Species", London, John Murray, 1859, p.189
23. Michael Denton, "Evolution: A Theory in Crisis," 1986, p. 250.
24. Michael Behe, "Darwin's Black Box: The Biochemical Challenge to Evolution", Free press, New York, 2006
25. Koonin, Eugene V., *The logic of chance: The nature and origin of biological evolution*. Pearson Education, NJ. 2012, p.351.
26. Video Clip of interview between Ben Stein and Dr Richard Dawkins, "Richard Dawkins Believes Extraterrestrials Created Man." https://www.youtube.com/watch?v=AiVoS78lNqM
27. Louis Pasteur, quoted in "The Literary Digest", 18 October, 1902.

5. Evidence 3: Intelligent Design

1. Walter Bradley, "The Just So Universe", in William A. Dembski and James M. Kushiner, "Signs of Intelligence", Baker Press, Grand Rapids, 2001, p.170
2. Robin Collins, in an interview with Lee Strobel, transcript in Lee Strobel, "The Case for a Creator", Zondervan, Grand Rapids, 2004, pp.161-162
3. William Lane Craig, "On Guard", David C. Cook, Ontario Canada, 2010, p.109
4. William Lane Craig, "On Guard", David C. Cook, Ontario Canada, 2010, p.109
5. Robin Collins, "The Teleological Argument", on http://citeseerx.ist.psu.edu/
6. Paul Davies, "The Mind of God", New York, Touchstone, 1992, p.16
7. 2002, Dear Professor Einstein: Albert Einstein's Letters to and from Children, Edited by Alice Calaprice, Page 127-129, Prometheus Books, Amherst, New York.
8. https://en.wikipedia.org/wiki/Proxima_Centauri_b
9. http://blogs.discovermagazine.com/d-brief/2016/02/22/earth-is-a-1-in-700-quintillion-kind-of-place/#.XFZp56ozZPY
10. https://blogs.scientificamerican.com/cross-check/can-faith-and-science-coexist
11. Isaac Newton, Principia, Book III; cited in; Newton's Philosophy of Nature: Selections from his writings, p. 42, ed. H.S. Thayer, Hafner Library of Classics, NY, 1953.
12. McGraw Hill Encyclopedia of Science and Technology. New York: McGraw Hill, 1997.
13. https://www.genome.gov/12011238/an-overview-of-the-human-genome-project/ See also, https://en.wikipedia.org/wiki/Human_Genome_Project.
14. See the Harvard University article, "The 99% of the Human Genome", http://sitn.hms.harvard.edu/flash/2012/issue127a/. See
15. https://www.nsf.gov/news/news_summ.jsp?cntn_id=118530
16. https://www.nsf.gov/news/news_summ.jsp?cntn_id=118530
17. https://www.ncbi.nlm.nih.gov/pmc/articles/PMC138976/
18. http://www.genomenewsnetwork.org/articles/02_01/Sizing_genomes.shtml
19. http://www.genomenewsnetwork.org/articles/02_01/Sizing_genomes.shtml
20. "Unlocking the Mystery of Life" DVD, Illustra Media, 2010
21. John F. Ashton. "Evolution Impossible", Green Forest, AR, Masterbooks, 2013, p.22.
22. Joseph Heller, "Catch 22", 1962, Random House, London
23. http://creationsciencehalloffame.org/defenses/design/irreducible-complexity/

24. https://profiles.nlm.nih.gov/SC/Views/Exhibit/narrative/doublehelix.html
25. https://www.telegraph.co.uk/technology/news/8789894/Monkeys-at-typewriters-close-to-reproducing-Shakespeare.html
26. Huxley used the term 'apes' but modern-day writers on this theme tend to prefer 'monkeys', e.g. David Osselton, 'Making a Monkey of Shakespeare', *New Scientist*, November 1, 1984, p. 39.
27. Dr Richard Dawkins, The Blind Watchmaker: Why the Eviodence of Evolution Reveals a Universe Without Design", 1986, Norton & Co, London.
28. Lee Strobel, "The Case for a Creator", Zondervan, Grand Rapids, 2004, p.275f
29. Miachel Ruse, in Lee Strobel, "The Case for a Creator", Zondervan, Grand Rapids, 2004, p.307
30. Michael Polanyi, "The Anatomy of Knowledge", Rutledge Press, London, 1966 p.323
31. https://www.gaia.com/article/consciousness-is-a-big-problem-for-science
32. J. P. Moreland, in Lee Strobel, "The Case for a Creator", Zondervan, Grand Rapids, 2004, p.307 f.
33. Roy Williams, God Actually, ABC Books, Sydney, 2008, p.69
34. Roy Williams, God Actually, ABC Books, Sydney, 2008, p.69
35. Quoted in Peter Watson, A Terrible Beauty: A Cultural History of the Twentieth Century, Phoenix Press, 2000, pp.673-4
36. Paul Davies, Are We Alone?, Penguin Books, 1995, p.84
37. John Polkinghorne, "God's Action in the World", J.K. Russell Fellowship Lecture, 1990, pp.3-4.
38. Albert Einstein, "Physics and Reality", Franklin Institute Journal, March 1936.
39. Roy Williams, "God Actually", ABC Books, 2008, p.68
40. John Polkinghorne, "God's Actions in the World", J.K. Russell Fellowship Lecture, 1990, p.4.
41. https://www.youtube.com/watch?v=of-8Q3HySjE&t=44m08s
42. Paul Davies, "Are We Alone?", Penguin Books, 1995, p.80.
43. Paul Davies, "The Mind of God", Simon & Schuster, 1992, p.220.
44. Paul Davies, God and the New Physics, New York, Simon and Schuster, 1983
45. From E. Salaman, "A Talk With Einstein," The Listener 54 (1955), pp. 370-371, quoted in Jammer, p. 123
46. Statement by Isaac Newton reported in Charles Dickens's "All the Year Round" (1864), Vol. 10, p. 346; later found in "The Book of the Hand" (1867) by A R. Craig, S. Low and Marston, p. 51:

6. Evidence 4: The Existence Of Objective Moral Values

1. Jeffrey Dahmer, interview on NBC Dateline, February, 1994
2. Jeffrey Dahmer, interview on NBC Dateline, February, 1994
3. William Lane Craig, "On Guard", David C. Cook, Colorado, 2010, p.137.
4. C.S. Lewis, "Mere Christianity", 1952, republished 1980, William Collins, London. Kindle Edition, p.2
5. See C.S. Lewis, "The Abolition of Man", Oxford University Press, 1943, republished 1978.
6. C.S. Lewis, "Mere Christianity", 1952, republished 1980, William Collins, London. Kindle Edition, p.2
7. C.S. Lewis, "Mere Christianity", 1952, republished 1980, William Collins, London. Kindle Edition, p.6
8. C.S. Lewis, "Mere Christianity", 1952, republished 1980, William Collins, London. Kindle Edition,p.2
9. William Lane Craig, "On Guard", David C. Cook, Colorado, 2010, p.142.
10. Michael Ruse, "Evolutionary Theory and Christian Ethics," in *The Darwinian Paradigm* (London: Routledge, 1989), pp. 262, 268-9
11. C.S. Lewis, "Mere Christianity", 1952, republished 1980, Harper Collins, London. Kindle Edition,p.11

7. Evidence 5: God's Intervention In Human History

1. Paul L. Maier, "Biblical Archaeology: Factual Evidence to Support the Historicity of the Bible", Christian Research Journal, volume 27, number 2 (2004)
2. Bryan Windle, "Biblical Sites: Three Ways to Date the Destruction of Jericho", in Bible Archaeology Report, May 17, 2019.
3. IBID
4. Kathleen M. Kenyon, "Excavations at Jericho", 3:110, London, British School of Archaeology in Jerusalem, 1981.
5. Neil Carter, quoted inpatheos.com/blogs/godlessindixie/2014
6. https://www.youtube.com/watch?v=u9CC7qNZkOE
7. John Shelby Spong, "Jesus for the Non-Religious", Harper Collins, 2007, p.69
8. "Myth Growth Rates and The Gospels", http://www.bibleinterp.com/articles/2013/kom378030.shtml
9. BT, Sanhedrin 43a, quoted in https://en.wikipedia.org/wiki/Yeshu#Yeshu_the_sorcerer

10. IBID
11. Flavius Josephus, Antiquities of the Jews, 18.3.3.63
12. Frank Morrison, "Who Moved the Stone?", Authentic Lifestyle, Great Britain, 1983, p.29
13. C.S. Lewis, "Mere Christianity", 1952, republished 1980, Harper Collins, London. Kindle Edition, p.32

8. Evidence 6: The Resurrection Of Jesus Christ

1. Plegon of Tralles, quoted by Origen of Alexandria (182-254 AD), in Against Celsus (Book II, Chap. XIV),
2. IBID
3. Flavius Josephus, "Antiquities of The Jews", 18.3.3 §63
4. "The Testimonium Flavianum", http://www.josephus.org/testimonium.htm
5. Luke T. Johnson, "The Real Jesus, San Francisco, Harper Press, 1996, p.136.
6. N. T. Wright, "The New Improved Jesus", Christianity Today, September 1993, p.26.
7. John A. T. Robinson, *The Human Face of God*, Philadelphia: Westminster, 1973, p. 131
8. Jakob Kremer, *Die Osterevangelien--Geschichten um Geschichte* ,Stuttgart: Katholisches Bibelwerk, 1977, pp. 49-50
9. D. H. Van Daalen, *The Real Resurrection*(London: Collins, 1972), p. 41
10. Gerd Lüdemann, What Really Happened to Jesus?, trans. John Bowden, Louisville, Kent.: Westminster John Knox Press, 1995, p. 80
11. Quoted in Josh McDowell, "Evidence for the Resurrection", Regal; Publishers, California, 2009, Ch. 16.
12. Joachim Kahl, "The Misery of Christianity", Penguin Books, 1971, p.119
13. reasonablefaith.org
14. Josh McDowell, "Evidence That Demands A Verdict", Thomas Nelson Inc, 1979
15. Frank Morrison, "Who Moved The Stone", Zondervan Press. 1930 (latest edition 2002)
16. Lee Strobel, "The Case For Christ", Zondervan Press, 1998
17. https://en.wikipedia.org/wiki/Simon_Greenleaf
18. C. S. Lewis, "Surprised By Joy", Harper Collins, New York, 1955, Ch. 14, p. 266

10. Pascal's Wager

1. https://www.brainyquote.com/quotes/winston_churchill_135270
2. Augustine of Hippo, "Confessions", written between 397-400 AD.
3. Augustine of Hippo, "Confessions", written between 397-400 AD.

www.ingramcontent.com/pod-product-compliance
Lightning Source LLC
Chambersburg PA
CBHW050310010526
44107CB00055B/2179